Intended for students, *Studying English* presents the ideas and debates that shape literary studies in America today. Why? Because if you know what's really going on "backstage", you understand more, find it more enjoyable and do better at it. The book covers arguments about criticism and theory, value, the canon, Shakespeare, authorial intention, figural language, narrative, writing, identity, politics and the skills that you can learn from English for the world of work.

In a clear and engaging way, Robert Eaglestone and Jonathan Beecher Field:

- orient you, by exploring what it is to study English in America now
- equip you, by explaining the key ideas and trends in English in context
- enable you to begin higher level study.

Studying English is the guide for students studying literature in the United States.

Robert Eaglestone is Professor of Contemporary Literature and Thought at Royal Holloway, University of London, UK.

Jonathan Beecher Field is Associate Professor of English at Clemson University, United States.

Studying English

A Guide for Literature Students

*Robert Eaglestone
with Jonathan Beecher Field*

LONDON AND NEW YORK

First published 2016
by Routledge
2 Park Square, Milton Park, Abingdon, Oxon OX14 4RN

and by Routledge
711 Third Avenue, New York, NY 10017

*Routledge is an imprint of the Taylor & Francis Group,
an informa business*

© 2016 Robert Eaglestone and Jonathan Beecher Field

The right of Robert Eaglestone and Jonathan Beecher Field to be identified as authors of this work has been asserted by them in accordance with sections 77 and 78 of the Copyright, Designs and Patents Act 1988.

All rights reserved. No part of this book may be reprinted or reproduced or utilised in any form or by any electronic, mechanical, or other means, now known or hereafter invented, including photocopying and recording, or in any information storage or retrieval system, without permission in writing from the publishers.

Trademark notice: Product or corporate names may be trademarks or registered trademarks, and are used only for identification and explanation without intent to infringe.

British Library Cataloguing-in-Publication Data
A catalogue record for this book is available from the British Library

Library of Congress Cataloging-in-Publication Data
Eaglestone, Robert, 1968– author.
Studying English : a guide for literature students / Robert Eaglestone ; with Jonathan Beecher Field.
pages cm
Includes bibliographical references and index.
1. English literature—Study and teaching (Higher)—Great Britain.
2. English literature—History and criticism—Theory, etc.
3. English literature—Outlines, syllabi, etc. I. Field, Jonathan Beecher. II. Title.
PR51.G7E256 2015
820.71'1—dc23
2015019459

ISBN: 978-0-415-83725-5 (hbk)
ISBN: 978-0-415-83726-2 (pbk)
ISBN: 978-0-203-38009-3 (ebk)

Typeset in Times
by Apex CoVantage, LLC

Contents

Illustrations — vii
A note for students — ix
A note for professors — xi
Acknowledgments — xv

Part I
How we read

1 Studying English — 3

2 Where did English come from? — 11

3 Studying English today — 23

4 English and disciplinary consciousness — 33

5 Critical attitudes — 43

Part II
What we read

6 Literature, value and the canon — 53

7 Castle Shakespeare	**65**

**Part III
Reading, writing and meaning**

8 The author is dead?	**83**
9 Metaphors and figures of speech	**95**
10 Narrative and closure	**103**
11 Creative writing and critical rewriting	**111**

**Part IV
English and you**

12 English, identity and politics	**121**
13 Why study English?	**133**
Conclusion: The importance of English	**143**
Further reading	149
Index	161

Illustrations

Figures

3.1	The older consensus	25
3.2	Different ways of interpreting	26
3.3	Studying English today	29
8.1	The traditional approach	84
8.2	After the death of the author	90

Table

5.1	Intrinsic and extrinsic critical attitudes	48

A note for students

Why English? You'll have your own reasons: perhaps you've always loved reading, thinking and talking about your favorite books and writers; perhaps you're attracted to the range and infinite variety of the subject; perhaps an English course is a requirement. You'll want to find your study stimulating and satisfying, and you'll want to get good grades. This book can help. Because if you know *why* a subject is the way it is, if you have a sense of the bigger picture, an idea of what's going on more widely, a subject is easier to understand, more enjoyable, and you become better at it. And this "why" is what *Studying English* is about. The book seeks to orient you in English by explaining some basic concepts, showing why they are important and by giving you the background knowledge to explore new ideas. It discusses the often unspoken links between courses you take and their relevance to wider debates about literature. And if you want to know what practical skills you're learning from studying English, that's here too.

Studying English does this by looking "behind the scenes" of the subject. The book will give you a sense of the arguments and developments that took place "backstage", and are still going on, and how these shape your course of study. This, too, will offer you some insight into why English faculty teaching and research interests are the way they are. We hope this backstage view will lead to different kinds of conversations and debates about literature and how it is understood. This might mean coming to appreciate multiple views of the same novel or poem or seeing why you may be required to take a course on Shakespeare for your English degree while your friend from high school majoring in English at another university does not.

Informed and engaged discussions of literature and ideas continue beyond college, and, of course, with people who did not choose the English major. The ability to discuss and debate words and ideas is not only important and pleasurable in its own right, but it's also crucial to the world of work now and in the future. More, while it is not a requirement for citizenship, this ability is necessary for an informed citizenship. The skills one learns as an English major are crucial for the future of democracy. So, when your engineer friends ask what you can do with a BA in English (and they will), you can not only point out the crucial skills you are learning (and perform them through argument!) but also show that you are playing a vital role for democracy.

Overall, we hope that this book will enrich your growth as a student of literature: helping you to get higher grades and, more significantly perhaps, to develop a deeper understanding and enjoyment of the work that you do in college. Please feel free to let us know how it worked out for you.

JBF & RE

A note for professors

This book was written to be of direct use to your teaching: indeed, *Studying English* had its origins in the experience of teaching. Students, reading literature, criticism and especially reading theory, and sometimes struggling, demanded to know *why* they were reading this or that particular text. I discovered (rather obviously, in retrospect) that when I explained *why* they were studying this text or topic, *why* this cluster of ideas was important, they found their study significantly easier, more enjoyable and they did better.

In exploring the "why", in trying to look backstage, *Studying English* openly addresses an awkward but well-recognised aspect of the subject. In some ways English seems outside the institution of education: the study of literature is about an array of things that are deeply significant but hard to pin down or test: personal response and experience, passion, interest, exploration, otherness, community, delight. In other ways, of course, as one of the most popular and often required academic subjects, it is very much within that institution. In education in general, there is a tension between the concrete objectives of obtaining certificates, degrees, skills and qualifications, on the one hand, and, on the other, a more intangible but no less real sense of personal and communal betterment. In English, with its complex relationship to the institution of education, this very real tension seems at its most obvious. "Will this be in the test?" can irritate the teacher, while the ambiguities of literary meaning and a lack of clear authoritative answers can irritate the student.

However, this book seeks to go beyond this tension by explaining to students why English is taught and studied the way it is: its focus is on what some educationalist call *metacognition*. Roughly, this means knowing what you are doing and why: knowing this not only makes you better at doing it, it makes it more rewarding and pleasurable too. Understanding the wider context helps focus the more specific learning which is taking place. Metacognition is crucial for making students informed, independent learners who can make connections and develop interests for themselves in their education.

Of course, this sort of overview of English is both challenging and problematic. There is little general consensus about the aims, approaches, purposes or even material to be studied: English characterised by dissensus, in fact. More, English is a fissiparous and quickly-moving academic field (some of our colleagues in other disciplines find this "faddish": but it might equally well be seen as "responsive" or even "responsible"). And students often take a range of electives with no central core and only limited links between those courses. But this book aims to take up precisely this challenge, in an introductory and not too programmatic way.

This is why the book spends a little time explaining how the history of the discipline shapes its palimpsestic, confusing and often contradictory present: how and why it became what it is now and what this actually means for students of English (chapters 2 and 3). The book then introduces the idea of an open-ended and fluid "disciplinary consciousness," describes how this is learned as a process (chapter 4) and then outlines some general critical attitudes (roughly, formalism and historicism, in chapter 5).

Central to the book is the idea of "thinking as a critic." Just as a mathematician (obviously) doesn't learn all the (infinite) answers to all the (infinite) mathematical problems but ways of thinking about and solving them, and just as a geographer learns to think about space and locations in certain specific ways, so English teaches students to think "as" critics. This may once have been but is no longer a sort of monolithic, fixed identity: rather it is a mobile, developing sense of a range of questions and ideas about the literary, widely defined, and, again, is characterised by dissensus. Learning to "think as a critic" is a process, which is why the second part of the book introduces longstanding debates and disagreements which have shaped the discipline and how it thinks: over value and the canon (chapter 6), understanding Shakespeare (chapter 7), authorial intention (chapter 8), figural language (chapter 9), narrative (chapter 10) and creative writing (chapter 11). In each case, the book explains why these are important and controversial. Arguments around these topics, often made implicitly, play a central part in English, and it's important that students know about them and what the stakes are.

A NOTE FOR PROFESSORS

The book stresses throughout how contentious English is as a subject, and chapter 12 addresses the often stormy relationship between English, identity and politics. The final chapter focusses directly and explicitly on debates over instrumentality in English and on the skills that English can teach students for the workplace: that is, it robustly answers the Gradgrindian question "What use is a major in English?" while showing why that question is itself contentious.

Studying English is designed to work as a primer or as pre-course reading, so that your students enter your seminars or lectures with the confidence and knowledge which allows them to understand and contextualise what they are studying and why. But it is also designed to be read as shorter pieces which can easily be integrated into your more focussed teaching. It can be applied to a variety of courses as it uses a broad variety of examples from literature and culture, and aims to be accessibly written. The book is short and the topics contentious. Tensions between your views and the text are inevitable but, perhaps, these may set the stage for your own interventions.

The discipline of English, of course, faces some challenges, although the situation is not as bleak as some paint it, perhaps. But at its best, it is a subject that can form responsible, knowledgeable, thoughtful, literate and highly-employable individuals who retain a life-long passion for literature and culture. This book, in introducing the subject to students and explaining why it is the way it is, seeks to play some part in this.

JBF & RE

Acknowledgments

By Robert Eaglestone

I've got a long list of people to thank for this book, but mostly, I am grateful to the students I have taught in English courses for showing me how to study English.

Lots of people have read drafts and proposals and made useful comments: Carol Atherton, Pamela Bickley, Barbara Bleiman, Mark Currie, Naomi Harbidge, Tim Kay, Caroline Mills, Jenny Stevens and Gary Snapper, Jennifer Neville, Adam Roberts and Sara Salih. I profited particularly from conversations about this book with Ewan Fernie, Judith Hawley, Douglas ("Are you ready for some criticism?") Cowie and with Kristen Kreider and Sophie Robinson. I'd also like to thank the following for their support: Debbie Wheeler, Sarah-Jane Duval-Hall, Lisa Dacunha and Laura Shoulder. I am extremely grateful to Polly Dodson for her editorial commitment and enthusiasm, to Ruth Hilsdon and to Deepti Agarwal. Thanks also go to her predecessors, Talia Rodgers and, especially, Liz Thompson, who edited earlier versions of this book with amazing understanding, passion and attention, above and beyond the call of duty. Errors are, of course, mine.

Thanks to the staff at the British Library (especially in Humanities 2) for being so unfailingly helpful. I also want to thank Jason Hodnett and Liz Allison for their assistance and advice on another matter during the completion of this book.

Reading and writing books is also a social process, so thanks to these people for their help and friendship: Eva Aldea, Shahidha Bari, Oli Belas, Matthew Broadbent, Penny Crawford, Holly Crocker, Tommy Crocker, Sarah

Dimmerlow, Mogs Eaglestone, William Eaglestone, Sarah Elgie, Finn Fordham, Malcolm Geere, Jane Geere, Simon Glendinning, Geraldine Glennon, Sophie Goldsworthy, Nicole Gyulay, Adrian Harvey, Nick Hoare, the Kelleys, Barry Langford, the Livseys, Sean Matthews, Martin McQuillan, Ankhi Mukhergee, Alex Murray, Hilary Sanders, Benjamin Poore, Danielle Sands, Helen Smith, Gavin Stewart, Dan Stone, Richard Tennant, Sarah Tennant and Julian Thomas, with a very special thanks to Poppy Corbett.

By Jonathan Beecher Field

I would like to thank Bob Eaglestone and Polly Dodson for giving me the chance to work with them on this project. It was very generous of Polly to imagine that my rambling observations on *Doing English* might qualify me to advise on this revision for U.S. readers. It was equally generous of Bob to engage so thoughtfully with my views on what his book ought to say.

Whatever I have contributed to this book, I owe to the Strode Tower of Power and the Clemson English students, faculty and staff, who make the Tower worthy of its nickname. If students (and parents) were not willing to take a chance on an undergraduate liberal arts major at a university better known for its engineers, I would have to find a new line of work, which would make me sad because I love my job. Some of the conversations with colleagues about what it means to study English have been over a pint at Nick's, while others took place during faculty meetings that threatened to end in tears. I am grateful for what I have learned from both kinds of conversations. A good bit of the thinking I've done about what it means to major in English happened during my term as director of undergraduate studies for Clemson English. I was extremely fortunate to work with an extraordinary administrator, Grace Ammons; a spectacular student services coordinator, Lindsey Kovach; and a miraculous fiscal analyst, Beverly Pressley. I am grateful for their underappreciated and undercompensated work.

Overlapping with these thanks is my gratitude to the many people – some real-life friends, some not – who have shared, over these last many years on social media, ideas about what it means to study English. Overlapping with these virtual communities is the very real community of the Society of Early Americanists. Our discussions of the particular challenges of teaching seventeenth-century sermons in the twenty-first-century classroom have sharpened my thinking about teaching in general.

Thanks to my family. I am fortunate to have Amy Monaghan as both partner and colleague.

PART I
HOW WE READ

1
Studying English

- Who is this book for?
- What is it for?
- How to use this book

Who is this book for?

This book is about *why* and *how* we study English. It aims to explain key ideas about the subject called English and the study of literature in general. If you are studying English literature either as a major or minor at a college or university or in AP classes in English in high school, this book is for you. In fact, whatever literature course you are taking, this book is not only an ideal stepping-stone to higher education but also an introduction to significant new questions and ideas about English and literature of all periods.

English often seems very different from other subjects. It isn't just that reading literature is (usually but not always!) pleasurable. More than that, in English, knowledge is made through the experience of reading and isn't simply passed down from authorities. And that form of knowledge, too, can't be easily explained. Knowing that a story moves you deeply is certainly a form of knowledge but a hard one to write about for a test. Indeed, the result of studying literature can be unpredictable, not least because, when we read, our own experiences and imaginations – our own lives and communities – are inevitably bought into the class or seminar. English, like the experience of literature, is bound up very closely with how we live, how we are with others, with ethics. It's a strange subject, then, partially within the systems of education (you have to pass tests) and partially outside (it's about who and how you are). This is part of the reason that students find themselves drawn to it (and, perhaps, that some don't like it). This book aims to explain why this is. Moreover, as I'll show later in the book and centrally in chapter 13, English teaches vital skills and broadens capacities for life and for work.

Even though English is probably the most popular arts and humanities subject, perhaps rather surprisingly there isn't a clear answer to the question, "What is English?" To say that it is the study of literature, analyzing writing or simply reading novels, poems and plays and then thinking and writing about them doesn't really answer the question. What does "learning about literature" or "studying English" actually mean? What ideas does it involve? Why do it one way rather than another? Why do it at all? People usually begin "studying English" without thinking about *what* they are doing in the first place and, perhaps more importantly, *why* they are doing it. The answers to these questions are vital because they shape what you actually do and how you react to the literature you study. And because the study of English is "half in and half out" of the normal processes of education, these questions are all the more complex.

Teachers of English at all levels in education have had long and tortuous debates (and even so-called "culture wars") over these questions – over what the subject is and how to study it – but these debates have rarely been explained to you, the person who is actually studying English, even though they affect your assessments, papers and projects, as well as what and even how you read. Some people think the ideas are too complex for students beginning to study the subject: I disagree. I think lots of questions about English (such as, "Is there a right answer?" or "Why are we doing this?" or "Why is it called English?") crop up right at the start. *Studying English* aims to explain these ideas and show how they influence you.

Why is it important to know about these ideas? John Hattie, an expert in education, undertook a huge "study of studies," covering some 80 million (!) students over many years. He argues that what he called "metacognition" – he means, roughly, "knowing what you are doing" – is crucial to improving a student's work. This makes sense: I believe that if you know *why* you are studying something, the subject becomes easier to understand, and you become better at it. In English, this means what helps you to do your best is not just knowing the texts but knowing *what* you are doing with them and *why*. Moreover, these ideas are interesting and important in their own right.

This book is shaped by four core ideas about English.

One: Reading is active

First and most important is the idea that reading is an *active process*. It can seem passive – you often do it sitting or lying down, after all – but it isn't a natural process; it doesn't just happen. Reading is a *dynamic act* of *interpretation*. And knowledge is made through the experience of reading and can't simply be "poured into you," as if it were water and you were a bucket. This means that "reading" and "interpreting" mean almost the same and you'll see I use the words almost as synonyms in this book.

When you interpret, it means that you find some things important and not others or that you focus on some ideas and questions and exclude others. You bring your ideas, your tendencies and your preferences – *yourself* – to reading a book, hearing a poem, seeing a play, watching TV or a film or looking at social media on a screen: your interpretation is shaped by a number of *presuppositions*. These are the taken-for-granted ideas and preferences you carry with you, and you always read through them, like glasses that you can't take off. On a surface level, your interpretation is affected by the *context* in which you read and by the *expectations* you have of the text. For example, if you read a novel about the Civil Rights Movement for a history project, you'll think about it differently than how you would if you read it for entertainment. At a deeper level, your view is shaped by your presuppositions about yourself, other people and the world, presuppositions you may take so much for granted that you might not even realize you have them. At this level, everyone has different presuppositions because – simply – everyone is different, to a greater or lesser degree, and have been shaped by different experiences. People from different backgrounds, sexes, sexualities, religions, classes and so on will be struck by different things in any text. And everything you have read and experienced previously affects how you interpret everything you read now. This idea can be summed up by saying that everyone is "located" in the world. Some people argue that your interpretations will always be constrained by these presuppositions; others think that you can escape them. Whichever is the case, you can think about and analyze them.

All this means that *no interpretation is neutral or objective* but has to be argued for and explained. And it means that *how* we read is as important as what we read because our presuppositions to a great degree shape the meanings we take from literature. Part of the aim of this book is to explore the impact of this rather obvious but often forgotten idea that texts are interpreted. This book also aims to make us think about our presuppositions and how they shape how we read.

It is because of the importance of interpretation that I have used the word "text" regularly throughout this book. Apart from being shorter to write than "novel, poem or play," it emphasizes that reading is an act of interpretation – texts are things that are interpreted. The word "text" also makes it clear that it's not only literature that is interpreted; so are people's actions, television, music and posts on social media, for example. News is interpreted both when it is watched, heard or read and when it is put together by journalists.

Two: English is a discipline

Second, and stemming from this first idea, is that, although English can seem as if it is just reading books, it is a subject or, more formally, a *discipline*. All educational disciplines and perhaps all forms of knowledge grew from very basic human activities. Chemistry grew from cooking and making clothes (dyes and so on). Geometry means "measuring the earth," vital for early farming societies. Creative writing and criticism both come from listening to stories and poems or watching dramas – interpreting texts – and then responding by asking questions and talking about them, as well as writing about them in different ways. Moreover, every discipline is made up of the questions it asks of the material it has chosen as its subject: originally, practical questions ("What to mix together to make red dye?"); then, slowly, more abstract questions ("How does the process of dyeing actually work? How do the different substances involved react to one another and change?"). Similarly, acts of interpretation lead pretty quickly to quite complicated questions, ideas and debates (including debating what might count as literature and what might count as an interpretation). These sorts of ideas have come, through complicated histories, to form the discipline and shape what we do in English today. I look at those histories in chapters 2 and 3 because, even though these ideas are often "below the surface" and rarely discussed with students, they still shape how English is taught and learned. It can be a bit of shock to think of reading and talking about books, and so about ourselves and others, in terms of being a discipline. But English is a discipline that has spent a long time thinking about precisely this: it is a discipline that thinks about its own nature as a discipline, precisely because it can appear not to be discipline. And as a discipline, English has all sorts of questions and ideas that it brings to the study of literature: this book explores some of these. And I'll argue in chapter 4 – and in the rest of this book – that studying English involves coming to know about these questions and ideas and how they might change our understanding of texts.

Three: English is controversial

People who practice the discipline of history are historians, and those who study biology are biologists: however, for reasons that the next two chapters will make clear, even the name for people who study English is controversial. Indeed, the third idea that shapes this book is that English is not only very popular but also a very controversial subject – controversial often because of its subject matter but controversial also as a discipline itself because it is woven into deep moral and political visions about who we are, how we should live and how we see the world and others. People with very different views on politics, morals, religion, education, history (and everything else!) have clashed time and time again over the subject of English, and these clashes have shaped the discipline and how we read in particular ways. To think about English and how we look at literature is to see a reflection of these clashes among ourselves and among our cultures. This idea is developed throughout the book, and part of the reason for this book is to explain why this subject is controversial.

Four: English is constantly changing

Finally, this book tries to show that English, as well as how we see literature, is constantly changing. All disciplines change over time: chemistry is very different now than how it was three hundred, one hundred or even fifty years ago. Moreover, disciplines are born, grow and die out over time. English is a relatively new subject: its modern form is only just over three or four generations old. It is also one of the most quickly evolving and developing subjects: indeed, the study of literature has been transformed radically in the last thirty years or so, as have the studies of literatures in other languages. One result of this has been that there can sometimes be a large gap – even a disconnection – between the way you study English at college and the way you study it in high school. This gap exists because there has been a huge influx of new ideas into the discipline of English: ideas about, for example, feminism and gender, sexuality, the mind and the body, race, globalization, the environment and the contemporary world, about the use of digital technology and about other art forms, as well as ideas drawn from all sorts of other disciplines. These new ways of thinking about literature have stimulated new forms of studying literature and even helped rediscover books, trends and authors that were previously passed over or ignored. These new ideas, often summed up as "literary theory," created this gap, and studying English today means having a sense of what these ideas are and, crucially, *why* they have arisen. Because of these new ideas, English as a subject has become much more wide-ranging and challenging, and these changes have affected all of us who study or teach English. This book's aim is not to explain in great detail all the new ideas that make up so-called literary theory but to explain *why* they are studied.

How to use this book

This book is for anyone who wants to know why they are studying English. It aims:

- *To orient you* by explaining what you are doing when you are studying English and why you are doing it.
- *To equip you* by explaining basic key ideas.
- *To encourage you* to explore new ways of studying English.

Often courses, exams and assessment seem to be more concerned with facts than with ideas: people focus on dates, for example, and not *why* things happen. But this is a book about ideas and should be read in that light. For example, although I mention various people throughout the book, what is important about them is not so much their names or dates as the *ideas* they have had and passed on to others.

The book has four parts:

- How we read (chapters 1–5)
- What we read (chapters 6–7)
- Reading, writing and meaning (chapters 8–11)
- English and you (chapters 12–13)

Each part contains chapters that explore one idea that is central for English today. The book finishes with a "Further reading" section, which is broken down by chapter. This final section shows you where the ideas covered in each chapter came from and where you can read about them in more detail.

Each chapter starts with a list of questions and finishes with a summary highlighting the main ideas covered. A couple of chapters also have diagrams that help clarify important ideas. The book is designed to be read in chapter order and gets more complex as it progresses. Because each chapter builds on the preceding one, you may prefer to read one chapter at a sitting and allow the ideas it raises to sink in before you start reading the next one. Or you may not.

Having outlined how the book works and what it's for, I will now turn to the first question. Where did the subject of English come from?

Summary

- This book is an introduction to ideas about English and literature.
- Reading is an active process: it is an act of interpretation.
- All interpretations have presuppositions. We can explore and analyze our presuppositions.
- English is a discipline, shaped by certain ideas and by its history.
- English is controversial because both literature and how we read literature are involved in debates about who we are, how we should live and how we see the world.
- Like any discipline, English is constantly changing.
- If you know why you are doing something, you are better at doing it. This book seeks to explain why we study English in the ways we do.

2

Where did English come from?

- What are disciplines?
- How did English develop?
- How do those ideas still shape English today?

Subjects that seem so taken for granted in school course catalogs or that simply crop up in the names of college departments didn't just appear. Instead, subjects, or, more technically, disciplines, are constructed over time and reflect the worldviews of those who construct them. Moreover, disciplines are not just ways of studying things that already exist: oddly, disciplines shape what they study as much as what they study shapes them. The discipline of English and the category of things it studies – what we now call literature – are no exception. Like every discipline and category, they developed through specific decisions, general trends and historical movements: indeed, studying English as we do now would seem very strange indeed to somebody from the early nineteenth century. In this chapter, I'll look at the intellectual, artistic and social forces that shaped the discipline of English. This is a complex story because the invention of a discipline is more diffuse and harder to pin down than the invention of a particular thing (the lightbulb or the airplane, say). English is not a simple, monolithic object but a family of linked, evolving concepts that don't fit together neatly *at all*. The evolution of these concepts is, of course, intimately involved with other ideas and historical events, so this story tells us not only about the subject but also about the changing ways in which people have seen the world. And like any complicated, widespread and demanding human activity, English comes from many different sources, and these histories and their effects, although they are rarely discussed, are not simply dead, in the past. They still shape what we do when we study English today.

English as a palimpsest

In the ancient and medieval world, material to write on, usually vellum made from animal skin, was prepared by scraping it to make it smooth: however, because such material was in short supply, these pages were often reused by scraping off the original writing to make the page smooth again. On reused pages, however, both the newer writing and – more faintly – the original writing are visible. This is called a *palimpsest* (palim-sest): it means, literally, "scraped again," one document that has (at least) two scripts on it. Today, a palimpsest has become a metaphor for thinking about how things are marked by different layers or processes. Sherlock Holmes, for example, in trying to find particular footprints on a wet and muddy track over which many people have walked, compares it to a fifteenth-century palimpsest he had been reading. Geologists describe landscapes as palimpsests, where an earlier form of rock formation is still visible despite many thousands of years of other physical processes "overwriting" it. And English today is like a palimpsest. Although it may seem as if it is one unified subject – it has one name, it is one period in the schedule at school, it is one department in the university – in fact, the discipline of English has been shaped by different conflicting layers of meaning and history.

Although what we call literature is as old as civilization, the formal study of literature is really very recent. Of course, people have always talked (and often written) about the stories they heard, the poems they read or the plays they saw. All literary works imply discussion and understanding because they call for interpretation. What is important here is how this became a *discipline*, a school and university subject. Some disciplines were invented long ago. Some scientists, for example, argue that people were doing what could be recognized as science in Egypt 2,000 years ago and that the basic principles of scientific investigation (experiment, observation and conclusion) were formulated by the English philosopher and scientist Francis Bacon (1561–1626) around 400 years ago. Philosophy, too, can claim to have started more than

2,500 years ago. In comparison, English, as we recognize it today, is a very new discipline. It started to emerge in the last decades of the nineteenth century but wasn't really established as a subject until the early decades of the twentieth century. Before the nineteenth century, there was no subject that corresponded to the discussions people had about Shakespeare or the letters they wrote to each other about the books or poetry they had read. In fact, no one had even really defined what the category of literature might contain: until relatively recently, the term included what we would call history, geography, linguistics, biography, philosophy, sociology, politics, science and much more.

Before English: The nineteenth century

In the nineteenth century, the closest thing to what we know as English – and it was still pretty distant – was the study of the classics. The classics were the ancient Greek and Roman plays, poems and texts from which American society drew a great deal of inspiration. The study of these were crucial in making one an educated gentleman. [And I do mean gentle*man* – women generally weren't allowed to study them. Indeed, in her *A Vindication of the Rights of Woman* (1792), the early British feminist Mary Wollstonecraft (1759–1797) argued that the right to study the classics was vital for women's equality.] It was taken for granted (although perhaps not by everybody) that literature in English was at best an imitation of the classics and at worst only a mildly pleasant diversion. It certainly wasn't worthy of study in the way that the classics were. But this idea was to change, and the university subjects that existed before English, like an earlier script, played a large role in that change.

The most important of the academic disciplines that underlay the development of English was philology. *Philology* (etymologically, it means, roughly, "love of words") is the study of language and of the historical origin and evolution of languages. In part, it grew from the earlier traditions of Biblical understanding and exegesis, and it was perhaps the dominant humanist subject in the eighteenth and nineteenth centuries. Because of the questions it asked and its historical approach, it was sometimes seen as threatening to established ideas. Philologists used a huge range of texts from the past – religious, historical, those we now call literary – to trace back the "family tree" of languages to their origins. (Indeed, it's been suggested that the origins of philology were in magic, inasmuch as people believed that the more ancient the language, the more power it had because it was closer to the language spoken by God, Adam and Eve in the Garden of Eden. This link between ancient languages and magic is why, for example, spells in books like the *Harry Potter* series are in versions of Latin.) But by the eighteenth century, philologists were less interested in the magic of wizards than in using language as way of tracing national and, for some, racial identity ("uncovering our cultural DNA", we might say today). In doing this, philologists often put together exhaustively edited scholarly editions to present literary and other works in their fullest and best versions. Although this study was less interested in what texts meant – in what made them "literary," for example – and more in how they fitted into a wider historical picture, philology helped establish the sense that works could be studied historically, that they formed part of a tradition and that this tradition was somehow connected to national or communal identities.

At the same time, universities also taught *rhetoric*, as they had for hundreds of years: rhetoric is the art of how to speak and write well and persuasively. People teaching rhetoric often used what we now think of as literary texts for models for good (or bad!) writing and as examples to work through. (Rhetoric also underlies the study of composition.) Like philology, the study of rhetoric was also less interested in the meaning of texts and their literary merits than in the use of language itself. Interestingly and in contrast to the nationalism that seemed inherent in philology, studies in rhetoric stressed how English was a language amalgamated from others (from Latin, Saxon, Old German, Norse and so on) and how different words for roughly the same things created complex shades of meaning.

In some ways, these two strands came together very early in courses in English taught at colleges of the University of London. For the first professors of English, the division between what we think of as English and history didn't exist, and the study of literature was the study of a historical period exemplified through its literature. Literature was taken to be the best way to come to know a historical period and, through that, the "mind of a people." This is one reason why English is (or was) called English (and not literary studies, for example): the idea was that by studying English, one comes to understand who the "English" are. As Ted Underwood argues in his book *Why Literary Periods Mattered*, at this time the "English curriculum was explicitly designed to foster national self-consciousness (expressed practically, as mastery of English style)." This inherent nationalism, which remains in the name of the subject, is also, of course, one reason the subject itself is controversial. But this very powerful idea – that somehow writing "opens up" the inner mind of a people and a period – had more than just academic consequences. At around the same time that courses in English were starting in London and spreading across the English-speaking world, this idea was being used in even more questionable ways.

During the first half of the nineteenth century, the British Empire was the dominant world power, and central to that Empire was the colonial occupation of India. At the start of the century, the British ruled India not directly but through a company, the East India Company, which had a complex contract, or charter, concerning trade and the exploitation of territory. (A fictionalized but equally rapacious version of the East India Company is also the villain in the *Pirates of the Caribbean* films.) This charter was agreed to by the British Parliament and renewed every twenty years. In 1813, Parliament renewed the charter but made a number of changes. They increased the East India Company's responsibility for the education of the Indian population and at the same time made it much harder for the Company to support the work of Christian missionaries and preachers. Previously, the East India Company had helped to convert the Indian population because the people in charge believed that Christian Indians would be more honest, more hard-working and more supportive of the Company's colonial exploitation. They thought that studying the Bible and Christianity made the population more moral, if "moral" is understood in the rather narrow sense of being in agreement with the principles of the Company. However, many people in London thought that persuading someone to become a Christian was quite risky. (Perhaps this was because converting someone involved asking her or him a lot of searching questions, which Christianity then claimed to answer: the last thing Britain and the East India Company wanted was for anybody to ask searching questions about anything, in case their regime itself came into question.) The upshot of this was that the East India Company had to devise another way of making sure that the native population would be eager to follow an "English way of life," at least closely enough to be good Company servants. The literature of England came to be seen as a mold of the English way of life, morals, taste and the English way of doing things. So the idea arose that Indians could be taught to be more English by teaching them English literature, which was seen as a way of "civilizing" the native population. This aimed, in the words of the British politician and administrator Thomas Babington Macaulay (1800–1859) in his *Minute on Indian Education* of 1835, to create a "class of persons, Indian in blood and colour, but English in taste, in opinions, in morals, and in intellect." In 1835, this idea was made law by the English Education Act, which officially made English the medium of instruction in Indian education and required the study of English literature. So the idea of a school and academic discipline called English that involved reading and writing about novels, plays and poems written in English was formed in India quite as much as in England. Again, this is also part of the reason the subject is called English. The idea that the study of English literature was a "civilizing force" remained very strong, and this was one of the ideas that formed the discipline and brought it back to Britain and across the Atlantic to the United States.

During the nineteenth century, internal struggles threatened to tear Britain apart. The Napoleonic wars had exhausted the country; a huge increase in population and the continuing industrialization of the country led both to domestic unrest and to the growth of enormous cities filled with poor workers. Those in power felt that Britain was being overrun by these "barbarians" and that anarchy or revolution was just around the corner. By educating the "British savages" in "civilized English" values, they hoped to maintain the political and social status quo. Many thinkers and reformers did feel that education was good in its own right, of course, but the hope of preventing revolution was certainly always in the background. Latin or Greek, considered the highest forms of "civilization," were assumed to be beyond the reach of most people in Britain, whereas novels, plays and poetry written in English were not. In effect, the study of English literature was brought back to Britain to "re-civilize the native savages." The schools inspector, poet and thinker Matthew Arnold (1822–1888) is one of the most famous of these "re-importers." In his best known book, *Culture and Anarchy* (1869), he wrote that culture – and he means mainly literary culture – would make "all men live in an atmosphere of sweetness and light." We might see this as a little simplistic, but it is a measure of the great hopes that were pinned on the study of literature.

These sorts of ideas were influential in the United States too: as part of the establishment of the nation and in response to the changing nature of the country and its population, and then to the Civil War (1861–1865), the idea grew up that literature and an education in literature in English could help society cohere. Writers like Ralph Waldo Emerson (1803–1882) and poets like Henry Longfellow (1807–1882) influenced and supported this idea, which became a source for the discipline of English. For them, literary texts were not simply pieces of dry historical evidence for tracing the development of the language or schoolroom sources to be drawn on for examples of how to write well. Instead, literature was in itself beautiful, moving; it taught about "how to live," in the broadest sense, as well as cultivating a way to be and a vision of the world. In doing this, it became assumed that some writers, centrally Shakespeare but others too, were better, more valuable and more important. These were critics especially interested in evaluating and appreciating literary works as *literary*. They were opposed by those, schooled in the rigors of philology, who judged that "thinking about how to live" sounded rather too vague and broad and an excuse to spend time reading "pretty writing" ("belles lettres") rather than being a demanding study.

English began, then, at the turbulent confluence of many different streams, each offering different reasons for the study of literature and different ways to undertake it. Some argued that English was simply a "weaker" form of classics; for others, it had to be a rigorous study of the history of language and its development; for some, it had to be a training in clear and effective communication; for others, it was a view into a national soul, mired (to modern eyes) in the history of imperial colonization in the UK, of slavery in the United States, and of racism; still others though it served a political need to ensure social cohesion; and some thought it was a vision of what was the best for human life. And all these different and contradictory scripts were written, as it were, on the same page. It's not surprising, then, that the early history of English, as the subject coalesced as a school and university subject, was made up by different factions attacking one another in wars of words over many, many years. But in this prehistory, some powerful ideas about the study of English emerged that are still with us. Literature is thought to be interwoven in some very profound and significant way with personal and communal identity. Literature and its study can be a shaping force on students, on how they write and who they think they are, and this force is involved with wider political and social forces, some positive and some less benign. Moreover, this assumption that literature is to be studied in historical periods is still very powerful. Literary study is involved with the study and even creation of traditions. These very substantial ideas are involved with all aspects of people's lives and societies. Because of this interweaving, it is no surprise that English has always been a profoundly controversial subject, one that is constantly in flux.

This sense of controversy and flux was evident in the name that these different practitioners of English gave themselves: some considered themselves philologists, some editors, some linguists, some rhetoricians, some literary historians, some literary critics, some simply writers. However, the next stage in the story was to make one of these titles the more significant.

English into the twentieth century: The New Criticism

William Stoner, the central character of the novel *Stoner* (1965) by John Williams (1922–1994), is a student of agriculture at the University of Missouri in the second decade of the twentieth century. He is taking a semester-long survey English course in his first year, and, in one class, they read one of Shakespeare's most famous sonnets, the seventy-third, which begins

> That time of year thou may'st in me behold
> When yellow leaves, or none, or few, do hang
> Upon those boughs which shake against the cold,
> Bare ruin'd choirs, where late the sweet birds sang.

In a situation some will recognize, the teacher is irritably (and rather unfairly) going around the class asking the silent students what the sonnet means. Stoner is put on the spot: what does it mean? But under pressure, something dawns on him. He begins by saying "it means," as if he is going to give a regular answer (it means this or that), but then, slowly and rather self-amazed, he repeats "it means": he has understood, in some very deep way, not simply what the sonnet might mean but, in a sudden moment, how poetry itself has meaning. Stoner, as his name implies, is a hard man to understand – indeed, he often doesn't understand himself – but this moment of insight leads him to abandon agriculture for English. This rather moving scene (moving, at least, for teachers and perhaps students of literature) embodies some important aspects of what was called the New Criticism. Stoner and his teacher are not interested in the historical context of the sonnet or in Shakespeare's life; they are not interested in the general questions of language; they are focused solely on the "words on the page" (a phrase that became a slogan) and what those words mean. It represents an act of critical appreciation, or what became known as close reading.

New Criticism, which dominated English in the United States in the middle half of the twentieth century and seems "natural" to many, began in part as a response to the carnage of World War I and a general concern about the rampant growth of technology and the so-called Machine Age. The poets, academics and writers behind New Criticism, like John Crowe Ransom (1888–1974), I. A. Richards (1893–1979) and Allen Tate (1899–1979), all influenced by the Nobel Laureate T. S. Eliot (1888–1965), believed that the study of literature would restore a sense of humanity to the world, that literature was a source of hope. They aimed to create a subject that would study literature in English in its own right, not just as a source of examples of how English was used in Shakespeare's time, say, or as pale imitations of ancient Greek and Roman works. As Louis Menand writes, Eliot's influential essay, "Tradition and the Individual Talent" (1919), which I'll discuss again in this chapter, basically asks:

What does a poet need to know? And the answer is: Poetry. The corollary to this is that the best way to understand poems is by their relation to other poems. This is the premise without which the enterprise of academic literary criticism would be unable to function.

This idea led to the (scientific-sounding) argument that (Menand again) "there is such a thing as specifically literary language, and that literary criticism provides an analytical toolbox for examining it." Exploring this idea, I. A. Richards, began what he called "practical criticism," presenting a poem to a class without the name of the author, the date or any other information and then letting the class respond and judge the poem as it stood by itself as an artwork.

As Ransom's influential essay "Criticism, Inc" (1937) argued, New Critics sought a form of reading that would be "scientific" in the sense that it would be "precise and systematic" and so almost objective and that would be "developed by the collective and sustained effort of learned persons" – professionals in criticism (thus the "inc," like a company) – undertaking a mutual, shared enterprise (as science is). For Ransom, literary critics are not like curators in museums, preserving works: nor are they like connoisseurs, specializing in a private appreciation of a work. Instead, they work together to create a shared critical understanding of literature. While historical and linguistic scholarship is important, the final aim for Ransom was not to read literary texts as examples of history but as works of *literature* with their own intrinsic value. Similarly, he argued that criticism should not be based on a personal response (he argued that "I found this moving" or "funny" is not a critical but a personal response) or on a moral or political position (for Ransom, if you think a work of literature is politically useful or morally offensive, this, too, is not a critical but a political or moral response). Instead, criticism arose from making a judgment based on the words on the page that you are reading. The critic Cleanth Brooks (1906–1994), in his book *The Well Wrought Urn* (1947), stressed the ways in which a literary text played with ambiguity and complexity in its own form, without recourse to other contexts or writing. These critics taught other critics and teachers, and this sense of what criticism was filtered through the whole education system.

This sort of criticism looked freer and more exciting than philology and more open than historical criticism. It allowed an interest in contemporary literature, which could be read just as "closely" as anything older, and it meant that the act of interpretation was itself something like a shared creative act. This form of criticism, too, had the (often unacknowledged) merit of being easier to teach: it didn't require a huge library with volumes explaining the historical background of a poet's work or an archive of original sources. It didn't require detailed knowledge of a foreign language. New Criticism developed what have become some of the strengths of English: shared dialogue, working together and attention to how words mean. And, because it was seeking to "humanize" a damaged world, it seemed very significant.

But it also had a range of hidden assumptions. The belief that the study of literature has a civilizing mission to humanize people and provide values that can't be obtained elsewhere in the modern world implies that there is a clear model of what the "human being" is supposed to be (and so, clearly, what the human is *not* supposed to be). Moreover, it seemed to imply (in the sense that criticism was a profession or, like a science, based on shared presuppositions) that there are deeper-seated ideas about identity and language that everyone doing criticism must share (or they would not be doing criticism). These notions led to the idea that literary judgment should attempt to be "objective": your personal gut response and views didn't count because it was the "literary-ness" of literary language to which you responded. As I'll suggest in chapter 4, this led – and still leads – to all sorts of torsions for students. These ideas seemed to take literature away from the individual reader and toward "professionals in criticism" while asking, oddly, for the reader to demonstrate "sensibility." This sensibility is what was supposed to be expressed "naturally" when a literary text is read, implying that every person must have something in her- or himself that is capable of being moved by reading literature and that English as a subject can draw out in order to improve this sensibility (but who could recognize everyone's form or expression of sensibility?). New Criticism also implied that the inherent value of a literary text could be separated from its context, its history and its politics. New Criticism suggested the idea that there is a "canon," or an authoritative list, of great literary works that everyone with sensibility should study and admire. Authors like Geoffrey Chaucer, William Shakespeare, Jane Austen, Herman Melville, Robert Frost and Henry James would be included in the canon.

New Criticism came to create a consensus about why and how literature was read and understood. These ideas came to dominate the discipline of English throughout most of the twentieth century (although not without a number of dissenters). From New Criticism, English took the idea of close reading and the sense that, as a subject, it was doing something critical, in all senses: critical in that coming to understand literature was important; critical in that it works through critical reading; and critical in that it was the work of critics. And so the dominant name for what someone who did English became "literary critic." However, as the next chapter shows, this idea, too, soon came to be challenged.

Summary

- Disciplines are constructed over time and reflect the worldviews of those who construct them.
- English is like a palimpsest, and how we study it today is shaped by the history of its development.
- One ancestor of English is philology, which studies the development of languages and texts over time. Another is taught rhetoric, the art of how to speak and write well and persuasively.
- English also grew from the idea that literature can shape and mold people, as well as from the ways that this idea was used, both positively and negatively. The study of literature came to be seen as a unifying social force.
- In the twentieth century, a powerful form of reading called New Criticism developed and came to dominate English, focused on reading the words on the page. Working as a sort of dialogue among critics, teacher and students, it sought to develop a shared, critical understanding of literature and a reader's sensibility. It valorized a critical community and helped to define a canon of great literary works. To some degree, it disregarded context and had an implicit idea of what the human being was supposed to be. Because of its dominance, it played a central role in shaping English.

3

Studying English today

- Why did English change?
- What do these changes mean?
- What is literary theory?
- What does this mean for you, studying English?

In the last chapter, I argued that the discipline of English was like a palimpsest with different and contradictory "scripts" coexisting on the same page. For example, some believed that literature illuminates the inner mind of a people or of an age, so history is central to the subject, while others thought literature teaches how to communicate well and convincingly regardless of historical context; others argued that the study of literature shapes people's ideas and values making them more individual; still others wanted literature to act as a social cohesive force; some felt that the study of literature teaches us all how to live and should be widely shared, while others thought that we should study the words on the page and become objective, professional critics. This array of conflicting ideas meant that English was characterized by wide-ranging arguments about what the subject might be and how it was to be done.

New Criticism had been the dominant script on the palimpsest of English during most of the mid-twentieth century – so much so that it was taken to be *the* way of doing criticism rather than a particular form of engagement with literature. It brought with it a range of practices (like close reading) and assumptions. Perhaps most significantly, with New Criticism, English was linked to the view that people need to be humanized and to be provided with values through the study of literature and that a work of literature was in language that had intrinsically literary characteristics. Looking at the development of English, it is clear that these ideas were intended, subtly but firmly, to force people into a single *mold*, or identity. If we all interpret texts the same way (or could be taught to interpret them the same way), we would all agree.

Changing world, changing English

However, the world in which we live now is not the same as those who shaped the subject. Nobel Peace Prize Laureate and reformer Jane Addams (1860–1935), in writing on Shakespeare and social unrest in 1896, argued that the "virtues of one generation are not sufficient for the next, any more than the accumulations of knowledge possessed by one age are adequate to the needs of another." It is not enough merely to preserve the good things and ethical ideals of the past: each generation has "to enlarge their application, to ennoble their conception and above all, to apply and adapt them to the peculiar problems" that people face in the present. She means that, even though we have to recognize what seemed good in the past, it is crucial to continue to develop and expand the good in the present in order to address the continually changing issues, challenges and ideas of our own times.

Our age has many challenges: where once certainties and definite answers seemed clear, we are aware that there are rarely simple solutions and final judgments. We are much less sure about many things that used to be taken for granted. And our sense of identity, of who "we" are, has changed. And as worldviews changed, so did our expectations of English (and, as I'll suggest later, English as a discipline both shaped and reflected some of those changes). In the last decades of the twentieth century, the dominance of New Criticism had faded, and English as a discipline was changing.

These changes are most clearly explained and explored by looking at the crucial issue of *interpretation*. In chapter 1, I argued that understanding literature isn't "natural," doesn't "just happen" but is the result of acts of interpretation. These acts of interpretation bring with them all sorts of presuppositions and taken-for-granted ideas. New Criticism took for granted the idea that English had a *civilizing mission*, that there was a sort of shared *objectivity*, neutral and disinterested, in reading that could, as a professional enterprise, make criticism like a science. Thus, it was critical and not personal; it presumed that a reader must demonstrate *sensibility*, a natural response that just happens when a text is read; it looked for the *intrinsic artistic worth* of a literary text and was less interested in its history or politics; it maintained an idea that there is a *canon* of great literary works that everyone should admire. This way of approaching literature implies that we *should* think, read and make judgments in the same way: if everyone were the same, there would be only one valid way of reading. If we were to express this in a diagram, it might look like the graphic in Figure 3.1.

However, these ideas about interpretation have been profoundly challenged. The idea of the *civilizing mission* looks more like a process of forcing people into a fixed pattern of values, ideas and opinions by making them interpret texts in the same way. The idea that one could make an objective judgment through close reading seems very questionable because every reader brings her or his own presuppositions to a text. This is not to say that we shouldn't read closely or that we can't question our own presuppositions but to admit that, unlike an experiment that should reproduce the same result each time it is performed, each reading by each individual produces a different result. It might be that there is no single "literary" language that the "science" of criticism could uncover. New Criticism too, aimed at being a shared professional enterprise that worked through dialogue: this looks, now, as if it simply excludes those others who might not be viewed as "professional." Worse, it also looks as if the idea of dialogue is secondary to the idea of the profession. The idea of *sensibility*, a natural response that just happens when a text is read, seems to rely on a natural response to literature, while the very fact that English is taught seems to confirm that such a natural response doesn't simply arise. Moreover, the idea of sensibility implies that if you are not moved by a certain work of literature, you have somehow failed. But who decides what should move us and in what way? "Sensibility"

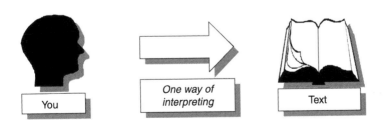

Figure 3.1 **The older consensus**

starts to sound a lot like "agreeing with me." The idea of *intrinsic artistic worth* begs the question, again, of who decided what that worth is and how it should be reckoned. The same is true of the idea of the canon of great literary works that everyone should admire: it implies that there are judgments of worth that could be neutral and disinterested.

New Criticism was challenged by a diverse range of new ways of reading which grew from many different sources. These new ways of reading are often grouped together under the catch-all term *literary theory*, although this isn't an ideal phrase. Because of the range of these ideas, their differences, and because these approaches rarely agree with each other, "literary theory" might better be put in the plural: "literary theories." However, theory is now understood to be central to English and the study of literature. Moving on from the idea that there was "one right way" of interpreting, these new approaches to literature reflect different concerns and ideas. Important and influential ideas have entered the subject: English now draws on history, politics, women's studies, sociology, gender studies, linguistics, philosophy and so on. And new ways of reading have also developed from within the subject of English itself, in reaction to the rather narrow focus of the New Critical approach to literature. This is why some professors don't like the term "literary critic" and prefer to be known as "literary theorists."

These changes are most clearly seen if we redraw Figure 3.1 to represent the new view of studying English (see Figure 3.2).

At the heart of literary theory, then, is the realization that every way of reading brings with it presuppositions. More than this, because everyone is

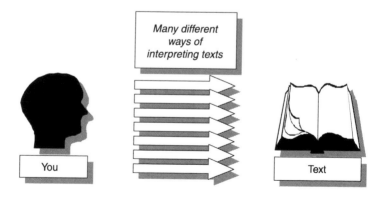

Figure 3.2 **Different ways of interpreting**

different, there simply cannot be one correct way of reading. But how is literary theory useful?

Using theory

This way of understanding what English is, as well as of understanding what you are doing when you are studying English, brings literature closer to you, the student. Those who shaped English at school, college and university have often not been clear about the ideas they take granted. Sometimes, these presuppositions might run counter to your own ideas and interpretations. Trying to work out or second-guess others unspoken and opaque ways of interpreting texts may mean that what you write in exams and papers about a work of literature might have nothing to do with what you really feel or think about it, which can be confusing and frustrating. To state this in formal terms, this marks the discrepancy between your location in the world and the older presuppositions of the discipline of English. Literary theory tries give more weight to different presuppositions and different ways of interpreting. Once you've realized that interpretations are determined by worldviews and that many interpretations are valid, you can begin to explore a wider array of ideas about literature. A key to this is remembering that you aren't limited to your own worldview: you can learn about different ways in which different people might interpret the same text. While your initial reading might be shaped by your presuppositions, literary theory offers a huge range of approaches to literature. You are free to choose one or another critical method, or to switch from one to the other, or to experiment with a selection. English becomes a question of reading certain sorts of texts in many different ways. There is no longer a *right* way to interpret literature. (I discuss this more in the next chapter.)

What are the actual mechanics of using different approaches to literature? Any critical method works by reading with certain questions in mind. The context in which we read, our expectations and experiences all make us concentrate on certain issues. These focus our reading and so structure our interpretations. For example, think about any of the novels, poems or plays you've studied. Now imagine you are asked any of these questions before you start to read: "What happens in the plot? Is this character likeable? How are metaphors being used to achieve a certain effect?" Each of these three basic questions will draw your attention to different parts of the text: the plot question will make you look at events, the character question makes you concentrate on what that character says and does, the question on metaphor makes you look at how the language is woven together. By focusing your attention on different aspects of the text, the questions make you read in a different way and so lead you to different interpretations of the text. You might even play down metaphor or plot if you are concentrating on character. But literary theories go beyond this and offer different sorts of questions to take into a text. Feminist approaches, for example, might suggest you ask, "How does this text represent gender and how it works?" Historical approaches might lead you to ask, "What is this text telling us about its historical period?" The text may or may not explicitly be about these things, but you make these questions your specific focus in reading and base your interpretation on them.

You can also think about the questions that shape other people's interpretations. If you're listening to a teacher or lecturer or reading somebody's thoughts on a work of literature, ask yourself, "What unspoken questions is she or he answering?" By uncovering these questions, you will learn a lot about that particular method of interpretation and about what that person thinks is really important. A greater challenge is to ask yourself what questions haven't been answered or haven't even been raised. Once you've worked through this, you can read the text with different questions in mind and see how different critical methods give different interpretations. Each will show up things the other methods don't.

There's no need, incidentally, to think that all these theories will agree with one another or add to a super-theory or a Grand Unifying Theory of Everything (and, in comparing English to a palimpsest, I've suggested that it never was one unified subject that "agreed with itself" anyway). In fact, the theories are more stimulating and productive when they don't agree. Indeed, to have lots of different critical approaches to texts means that we can compare and contrast them. If English is about reading texts in different sorts of ways, it is also about examining how and why we choose these ways. English is not only about *reading* and enjoying literature; it's also a question of *thinking about how we read*. We can show this on our diagram, by adding another arrow representing a focus on interpretation itself. The name for this study of interpretation is *hermeneutics*, which is what I've called the arrow in Figure 3.3.

The realization that *how we read is as important as what we read* is perhaps the most important innovation in the study of literature in the last thirty years. It has changed English completely as a subject and given it a new burst of life. And it is this realization that underlies the new ways of reading that are called, in a rather all-inclusive way, literary theory.

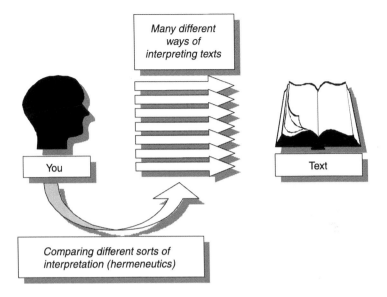

Figure 3.3 **Studying English today**

English today

Studying English involves not just reading works of literature but also learning to interpret them in different ways. It involves understanding how different interpretations work, revealing what other people consider to be significant about literature and central to their lives. Consciously reading from different perspectives can make you reflect on your ideas about texts, about yourself and about how you see the world. Many critics and educators say that this sort of questioning and reading from other perspectives is central to English and to the enjoyment of reading. I think that this power to make us reflect on ourselves and others is one of the reasons that English is such such a valuable subject and why literary theory is essential for English.

Change rarely comes easily. The issues raised by literary theory have caused terrible arguments and divisions between students and teachers of English in schools, colleges and universities, as well as very heated public debates. (In the 1980s and 1990s, these were part of what was rather melodramatically called the culture wars.) In part, this led to the gap between English in secondary and higher education that I discussed earlier. As a discipline develops, it becomes more complex and encourages more involved debates: indeed, some argue that a discipline becomes "fully mature" only when it questions the very criteria, aims and approaches with which it began. So English might begin by studying literature but then comes to ask, "What is literature, anyway? Why are we studying it this way rather than that way?" These are the more reflective questions that literary theory asks.

A discipline doesn't stand still. The early years of the twenty-first century saw what many thought was an increased interest in the historical contexts for understanding works of literature (some called this a "turn to history"). Others, too, argued that the day of literary theory – or of certain sorts of literary theory – was over: and certainly theory, like everything else in the discipline, changes and develops. In contrast, in 2014 Vincent Leitch, editor of the *Norton Anthology of Theory and Criticism*, argued that there is a "theory renaissance": many new forms of thinking about literature and culture, a fascinating and dazzling range of topics and ideas, all responding to the world we inhabit. And, of course, other strands of the discipline's complex past can become popular again: some faculty in English have been discussing, for example, a return to philology, whereas others seek to make the discipline more like versions of rhetoric.

Sometimes this can seem rather overwhelming or, because there is no one definition, threatening or unnerving. But I'm not sure it need be. Activities we undertake need not have one overriding aim: we can do a sport, for example, because we want to keep fit, because we enjoy the company or (and!) the competition and because we learn wider lessons from our dedication to practicing it. Adapting a metaphor from the philosopher Ludwig Wittgenstein, we might see English as a rope, made of twisting together many strands and fibers. Wittgenstein points out that the "strength of the thread does not reside in the fact that some one fibre runs through its whole length, but in the overlapping of many fibres." English as a subject is made up of many of these fibers, with new ones being added all the time. It does not need one, single thread running through it (even if this were possible). Further, these theories and questions arise precisely from the literary material that English studies: the stories we tell ourselves, the way we represent ourselves, are supposed to be communal and are supposed to involve us in our conversations about ourselves, others and our world. These are often controversial: that the discipline exploring these is also controversial should not be a source of worry. Perhaps, more encouragingly, the whole discipline is, itself, like a larger version of a lively class discussion of a literary text, wherein everyone offers different interpretations. The next chapter explores in more detail what these changes and ideas mean for studying English, along with how you can learn about them.

Summary

- Studying English involves reading works of literature, learning to interpret them in different ways and understanding how these different approaches work. These new approaches have the potential to create new readings of texts and to make you think about the way you see the world and your place in it. You are, or should be, free to choose one or another method or to experiment with a selection.
- English as a subject is constantly changing. These changes can be understood by looking at the issue of interpretation. When you read, you interpret. No interpretation is neutral or objective because we are all influenced by a number of presuppositions.
- "Literary theory" is a catchall term for a huge range of new and different ways to read and interpret texts, reflecting the different concerns and ideas of a very wide range of people, not just an elite. All this encourages us to think about how we interpret.
- Theory also encourages us to contrast and study different methods of interpretation in their own right. This is called hermeneutics, the study of interpretation itself.
- How we read is as important as what we read.

4
English and disciplinary consciousness

- How do these changes affect your study of English?
- What is disciplinary consciousness?
- Is there a right answer in English?
- How do you learn a disciplinary consciousness?

What do the changes I have discussed in the previous chapter mean for how we study English as a discipline? Students studying English are sometimes confused by the fact that different critics seem not only to disagree about what a work of literature means or about how good it is but also about how to approach it, and even the overall point of what it is they are doing. As I've suggested, the confusion isn't the student's fault but stems from more, perhaps, from the different practices in the discipline. This chapter suggests that thinking about English as a discipline, with its many different strands, helps resolve some of these difficulties.

Further, this chapter aims to explain what it's like starting to study a subject that is as much a palimpsest as English. In his book *The Humanities and the Dream of America*, Geoffrey Galt Harpham writes that in

> a great many colleges and universities, there is no "Intro to English" class at all, because there is no agreement among the faculty on what constitutes a proper introduction to a field in which the goals, methods,

basic concepts and even objects are so loosely defined, and in which individual subjective experience plays such a large part.

Ask your favorite professors about the last time your department changed the requirements for your English major, and they are likely to blanch and shudder at the memory of that trauma. The previous chapters have explained why this is the case: this chapter aims to show how students can engage with English even if it lacks a simply defined core.

What is a discipline anyway?

When you study math, you don't learn by heart the answers to all possible mathematical problems: that would obviously be impossible and mad. Instead, you learn methods to solve problems. Similarly, in geography, you don't learn everything on a map: you learn to think like a geographer, to ask the sorts of questions a geographer might ask. In education in general, you learn not so much facts or things by rote as *ways of thinking* about things in the world. You learn to think *as a* geographer or *as a* mathematician. English is so controversial that, as I've suggested, even the name of what it is you are learning to think as is complicated: any term (say, "literary critic," "student of literature," "theorist," "philologist") is freighted with expectations about what one should and shouldn't do and about what English is. However, for ease – and aware that it's contentious – I am going to use the term "literary critic." So, in English, you are being taught to think as a literary critic. Learning to think as a literary critic is, like learning any discipline, learning a *disciplinary consciousness*. Indeed, really, the whole point of this book is to introduce explicitly the disciplinary consciousness for English and for literary studies. Having this consciousness will empower you to have different and better conversations with your professors and classmates.

I'm going to suggest three ways to think about disciplinary consciousness. The first is to think about the questions a discipline asks. In chapter 1, I suggested that all educational disciplines grew from very basic human activities and developed from more straightforward questions to more complex ones. Chemistry developed from practical questions (such as how to cook) to more abstract ones (how does this process occur?). Literary criticism comes from reading stories and poems or watching dramas (and now films, television, social media and so on) and then thinking about them, asking questions and talking about them. The basic questions one starts off with are – like all proper questions – usually the sort that look simple but turn out not to be. As I'll show later in chapter 6, the question "What is literature?" becomes more and more difficult to answer the more one thinks about it. Interestingly, while it might seem as if the world is just there and disciplines simply "divide it up" into sections for study, there is considerable interplay between a discipline and what it studies: the questions one asks shape and define the area you study, as much as the areas you study pose certain questions. "What is literature?" is a case in point. As disciplines grew around it, more and more things were defined as "not literary": works of history or psychology, for example, rarely count as literature today (as they used to), and, as I'll argue in more detail in chapter 6, even the term "literature" started to take on a sense of evaluation, which meant that even some novels were thought not to be literature. So one way to think about the disciplinary consciousness of English is to discover what questions literary critics ask. (You have come close to understanding a discipline when you know what questions its practitioners ask and how they are answered.) What are these questions? Sadly, there isn't a simple checklist (although, if there were, wouldn't that be a very limited sort of discipline?): however, we can see that the questions we ask of literary texts begin with the same questions we ask every day of ourselves and of the people around us, although they soon become more involved, abstract and complex. Every day, we use language to express ourselves, tell stories and make patterns out of our reality. That means we judge, shape and think about language all the time, so much so that we often just forget we are doing it. Thinking as a critic both reminds us of that intimate involvement with language and, more importantly, develops out of that involvement. This is why there is a strange closeness among literary texts, language, criticism and ourselves, and that's why English can be so risky. When we read or talk about a text, we are risking ourselves, risking revealing ourselves. That's also why there are rarely straightforward answers: thinking as a literary critic is more about the process of asking questions and a growing awareness of how and why these questions are hard to answer.

Another way to think of disciplinary consciousness is, as Ben Knights and others argue, that it is like a "script," or a way of behaving and responding. A discipline is not only about its formal questions, its reading and its assessments but a whole range of the other formal and informal ways of thinking and behaving that students learn. Sometimes these are in the classroom or lecture hall: sometimes they are outside and serve to define one major as opposed to another. These can become cultural codes that shape the identity of the student (and the teacher) and, whether in the seminar or outside, can be quite influential.

Finally, disciplinary consciousness can be understood as a form of tradition. The idea of tradition brings out different reactions in people. Some are keen to embrace and inhabit traditions; others are keen to get rid of them and begin anew.

However, both extremes have dangers: too tight an embrace means that a tradition can become simply an ossified habit with no real meaning. In contrast, without responding to and facing the past, thinking in the present simply has nothing to go on (and risks making the same errors over and over again). In fact, all thinking and education take place in the context of engaging with the past, recent or distant, whether it moves beyond a tradition or reaffirms it. Moreover, a living tradition is characterized precisely by debates about the very point and activity of that tradition. Coming to a disciplinary consciousness means precisely this: coming to terms with the traditions – in the case of English, the controversial and contrary traditions – that have shaped the discipline. Indeed, being explicit about those traditions, as well as up front about their flaws and benefits, is part of what makes a discipline self-reflexive and mature.

One thing that's weird about English and really worth noting is that at the heart of it is the experience of engaging deeply with art and this experience is supposed to be both significant and pleasurable. It's why so many English majors are enthusiastic about their subject. And more, it's something that people do every day without being literary critics (or whatever term we might use). Once we start to "study English," however, the texts can sometimes become like chores or homework, as if they have lost that moment of importance and joy. It is as if taking on the disciplinary consciousness can, at first, get in the way of the art. Sometimes people say that analyzing a poem kills it. But, oddly, no one ever says, "You are murdering this landscape by doing geography all over it" or "Math is butchering my enjoyment of numbers." The gamble of English is that thinking about texts *as a* literary critic, when one has a sense of what this means, makes the texts more enjoyable, somehow more profound and interesting, not less (although it may be the case that learning to think as a critic, using disciplinary consciousness, is challenging at first). Again, part of the point of this book is to try to make this process easier. One of my proudest teaching moments was with a student who rather tentatively said to me, "I really enjoyed this novel . . . even though I had to write an essay about it."

Disciplinary consciousness and problems in English

Is there a right answer?

Thinking about English in this way, as a discipline, offers answers to a number of common student questions and confusions. One of the most important of these is the totally fair question, "Is there a 'right answer'?" Although it's often said that there is no right answer in English, papers are graded, authorities on literature are deferred to, and tests are revised, so it looks as if everybody secretly assumes that there really is a right answer and a way to get the best grades. Why are students and teachers of English caught in this contradiction? And what are the consequences?

Focusing on the right answer can lead to all sorts of problems. For example, it can lead to the idea that an author choses a theme on which to write and then decorates it to make it interesting. The job of the student is, then, to uncover the theme, like cracking a code. (What if there was no simple message or moral in the text?) Moreover, the sense of a right answer might mean that "we" are all forced to feel that we must enjoy the book and that we must also find it a masterpiece. This idea of "we" becomes not inclusive but coercive. It can create an internal split between the person reading the book (who might not like it) and the person who is supposed to be, say, writing a paper on the book (who has to pretend to like it). A symptom of this torsion is that students often write, "The play is flawed because . . ." or "This character is engaging because . . ." as a way of expressing "I don't like this" or "I like this character." Indeed, writing in the third person ("he/she/the reader") rather than the first ("I") is assumed to be more objective – while students are often told that their informed personal ("I") response is what is being sought. Further, the idea of the "we" suggests that literary texts have only one meaning and that disagreements are ways to uncover that one meaning rather than being different ways of interpreting that same text. This also means that a novel, poem or play can be reduced to the answers in SparkNotes or on Wikipedia.

But perhaps the most insidious problem that searching for the "right answer" creates is that some books become "course texts" while other texts, which you might read or enjoy in your spare time, do not have the same status. Course texts seem to have to be read – interpreted – in the same way as teachers and examiners, sharing their presuppositions (no wonder these course books often seem hard to read); the others don't seem to ask for this sort of reading. This split between what English sometimes claims to be doing (inviting you to respond to a text) and what it actually does (teaching you to respond in one particular way, corresponding to one set of presuppositions) has a number of consequences. Perhaps most importantly, this "split" makes a personal response impossible and often discounts a student's own experience, own reading, own ideas and presuppositions as "wrong." Students can become inhibited in their responses because they discover a hidden agenda. Indeed, students sometimes learn to speak two different "languages": one about their course texts and one about the books and poems read, the films and TV watched, the websites visited, the blogs written and read, the music listened and danced to in their free time. Precisely because of these conflicting messages, the subject seems much harder than it should be, and sometimes the only way students can resolve these messages is to find English either impossible or simply irrelevant to their own experience.

It's often said that, "You can say what you want as long as you can argue for it," and this is true to some extent. However, seeking the right answer often gets in the way even of this.

First, in learning to support a nonconventional argument, you need material and ideas that are rarely provided precisely because they aren't conventional. For example, you would find it hard to write an essay offering a psychological interpretation of a text if you had never been told what such an approach might involve or even that such an approach existed. Second, the way you are taught to argue relies on presuppositions, which you end up taking onboard perhaps without even realizing you are doing so: so how you argue and what counts as a good argument are already "set up." Is it a surprise that class after class of students in the end had the same opinion of Shakespeare's *Othello* if they were all surreptitiously drilled in the same basic ideas?

But the idea of the "right answer" in English is bad not only for students but also for teachers. Many teachers – rightly, I think – treasure the idea that English is not like other subjects: it aims to encourage freedom, personal responses and understanding. Perhaps even more ambitiously, some teachers have felt that English as a subject has a mission to make people more free and fulfilled in their lives and choices. Yet they have often felt confined by assumptions from the middle of the twentieth century, preserved in educational formaldehyde. Worse than just memorizing facts, to be good at English, students were forced to take for granted a very different set of attitudes and ideas. Teaching events like videos, summer schools, YouTube clips, performances, visits by actors and so on often showed how texts could be exciting and relevant. But if the bottom line was the grade, then much of this was wasted. This left many teachers frustrated and angry. So it looks as though the search for the right answer in English – and behind it the perfectly right search for good grades – leads to all sorts of problems.

The way to resolve this problem is to think about learning the disciplinary consciousness as a *process*. I said it would be mad and impossible to simply learn all the answers to all mathematical problems: instead, you learn a method, a way of thinking. The same is true for English, although it can be a longer process. Two British educators, Barbara Bleiman and Lucy Webster, suggest that in English there are six steps. (In real life, these all ooze into one another, so breaking it down like this is just to clarify the process.)

First, you begin with your own initial response to the text. This could be enthusiastic or interested or even negative (even being bored is an interesting response: why didn't the text grab you?): whichever it is, you should think about why you responded as you did.

Second, it's important to listen carefully to the responses of your friends and other people in your group. Somebody else may have seen something that you missed or may have a very different take on the text. When we read Mary Shelley's novel *Frankenstein* (1818) at university, I felt sorry for the (poor) monster: my classmates didn't at all. Criticism you read can be a bit like this too: different points of view that make you think.

Third, you could respond to "soundbites" about the text: short ideas or suggestions that illuminate it or suggestions from your teacher. You might come up with some initial ideas by looking at an online guide to the author or a relevant critical essay. Actually, titles for papers are very often soundbites that exist to make you think. So a good soundbite for *Frankenstein* might be from the ancient Greek philosopher, Aristotle: "to live outside society you must be a god or a beast." Victor Frankenstein, who makes the monster alone in his lab away from everyone, is like a god, creating new life: the monster, also alone and isolated, becomes a beast.

The fourth step would be to read critical essays or reviews about the text. Just like you, these critics have ideas and points of view that guide their responses to the text, and you can begin to ask what "position" each critic is writing from, what ideas underlie their work. For example, one critic, Franco Moretti (b. 1950), writes that the

> monster is denied a name and an individuality. He is the Frankenstein monster; he belongs wholly to his creator (just as one can speak of "a Ford worker"). Like the workers, he is a *collective* and *artificial* creature. He is not found in nature, but built.

Moretti points out that the monster is literally made of the bodies of the poor, reshaped and given an identity by modern science and industry. Moretti is obviously a critic with an interest in how politics and history shape a literary work and its contexts.

In the fifth stage, rather than reading critics who develop their own ideas, you could yourself explore different critical positions on the text (you could think of yourself as a feminist critic, say, or someone who is interested in the ethical or political issues in a text). Arguing from a certain position can be very revealing, and it also involves discovering something about the critical approaches themselves. (You will probably have begun to think about these approaches in the fourth step.)

The final step is when you "own" a critical position, when a set of ideas becomes part of your literary-critical tool kit, always ready to hand. This sixth stage, leads back, of course, to the first stage, only now your initial response will be much more informed and analytic. (Indeed, another reason why there is no right answer is that there is no real end to this constantly developing and circular process.)

These six stages also show that the disciplinary consciousness of English is developed as a process by studying particular texts and ideas and then reflecting back on wider concepts and assumptions, moving between the general and the particular. Understanding English, then, is not about knowing the right answer: it's about thinking as a literary critic.

Why is there jargon in English?

"Jargon" is a word you will hear people outside our discipline (and some inside) use to criticize the way we talk about the work we do. However, the idea of disciplinary consciousness in English also helps to explain why technical terminology or even jargon is used. There are three sorts of technical terms. The first sort are simply descriptive. It's easier to say "iambic verse" than "verse that goes du-dum du-dum du-dum du-dum." The second sort is made up of terms that implicitly present a wider critical approach. Each set of critical terms brings with it its own presuppositions and ideas, and many approaches use many different terms (although some do overlap). In general, these terms actually work in the same way as the more descriptive terms (it is easier, say, to use "reification" or "objectification" – after explaining them, of course – than to repeat "the processes of thought that turn people and abstract concepts into things"). Technical terms can be off-putting, but simply to oppose technical terms or the complex ideas about literature that they embody (as some people do) assumes that there could be a "natural" way of interpreting that does without a technical language of any sort. It also assumes that doing English should be easier than a subject like chemistry or sociology, where technical terms abound. The third sort of technical terms is perhaps the most interesting: sometimes jargony words or phrases can be the sedimented forms of those questions and debates that have shaped the field (as in, say, the "words on the page" in chapter 2 or, as you will see, the "death of the author" in chapter 8). In part, learning a disciplinary consciousness means learning what these sedimented forms mean and what they really are, seeing that why those are the questions to be asked.

You don't have to study English to read books, so what makes English special?

Disciplinary consciousness also speaks to another issue that is often raised in relation to English. You don't have to be an English student to read novels or to have a critical opinion about a film, so what makes English special or different from normal, everyday conversations about books, films or other texts? It's certainly the case that people who aren't English students know about Fitzgerald or Shakespeare and have illuminating things to say about them. Some people have found this a threat: as I suggested, the New Critics wanted a cadre of "professional critics" to work on literature. English certainly can give precision, detail and complexity to analyzing texts. Importantly, it also offers informed and different ways to interpret texts, which can be very revealing of the texts, yourself and others. It's also honest to admit that there are all sorts of different expertises on texts and that the disciplinary consciousness of English represents one way – or, more accurately, an interweaving of many ways – of coming to understand literature and other texts. The edges of the subject are porous and often very open. (English also teaches an impressive number of skills for the workplace: I discuss these in chapter 13.) However, it is also true that the idea of disciplinary consciousness seems to be a way to explore or even resolve some of the problems that people have in approaching art. For example, in newspapers or online, journalists find it easier to gossip about the writer's lives than discuss the work that has made the writer interesting in the first place. In other areas – in the world of computer games or in science fiction fandom – extraordinarily combative and painful arguments about gender, race and class are repeating arguments had in English over the years. By exploring these arguments in English, we can discover what is at stake in these other areas and why.

Conclusion

Disciplinary consciousness represents the idea of metacognition, which I mentioned in chapter 1: knowing not just the content of a subject but also what you are doing and why. As the world becomes more information rich – you can discover a fact instantaneously – knowing why a fact is important and what it means is even more significant. More, this "knowing why" leads as the educationalist John Hattie argues, to better grades. But the most important reason to think about the disciplinary consciousness of English is that it allows us to think and talk not only about texts but also how we come to understand and interpret texts. It helps dialogue and discussion: if it does not generate consensus – and perhaps we might be suspicious of too much consensus – it does at least create informed dissensus.

Summary

- A discipline involves "thinking as a . . .": English involves thinking as a literary critic.
- Thinking about disciplinary consciousness helps resolve problems found in studying English, such as, "Is there a 'right answer'? Why are there technical terms?"
- We learn the disciplinary consciousness of English as a *process*.

5

Critical attitudes

- Where should we start with thinking about how we read?
- What is the intrinsic attitude?
- What is the extrinsic attitude?

English can appear to be quite daunting once you realize that there's a near infinite number of ways you can read. If you're told to explore different methods of interpretation, to challenge your presuppositions and to think about how you read, where are you supposed to start?

In the last chapter, I suggested that learning about different critical approaches or theories is a *process* that you go through. A step in this process is to look for patterns in the way these critical approaches work. To do this is to look for the presuppositions behind them and to think about the contexts in which texts are understood. In this chapter, I shall outline one pattern that can be used as a starting point for thinking about a wide variety of critical approaches.

Into the text or out from the text?

If you look at a painting, are you looking through a window to another world, or are you simply looking at the composition of color and shape on a flat canvas? If you see a painting as a *window*, you might be concerned with what is going on behind the window: who the people are, say, and why they had their picture painted. You might ask about the historical significance of, for example, the skull on the shelf or even why the painter chose that particular subject in the first place. If, however, a picture is only a flat canvas, then you would ask different questions: about how the tones contrast or how the shapes relate to one another. You might just be struck by the beautiful range of colors.

This same contrast occurs in thinking about literature. When you read a novel, poem or play, how do you approach it? Do you look at it as a beautifully woven fabric of language? Or as an example of writing that tells you about the historical period in which it was written? Is it stimulating because it puts words together in a new way? Or because it pours out on paper the intense experiences and interesting ideas of a particular writer? When we study English, do we study literary works for their pure artistic merit or because they reveal things about the world and their authors? Do you think of yourself as going *into* the text for itself or coming *out from* the text to explore other issues?

One of the longest debates in English has been about whether interpretation should focus on the text as a text itself (a flat canvas) or on the text as evidence for something else, such as its historical period and its attitudes or an author's life (a window on a world). In an influential book called *Theory of Literature*, published as long ago as 1949, two critics, René Wellek and Austin Warren, called these two contrasting positions the *intrinsic* and *extrinsic* approaches to literature. These two terms are not the names for critical approaches themselves – instead they name contrasting sorts of presuppositions, tendencies or *attitudes* taken by approaches to literary texts. This debate, because it discusses what happens when we interpret in different ways and compares different methods of interpretation, is an example of hermeneutics, the study of interpretation. Certainly the debate has become more complex since 1949, but intrinsic and extrinsic attitudes are a good place to start.

Intrinsic attitudes: Into the text

The intrinsic attitude is often called formalism because it is concerned, above all else, with the *form* of the text, its structure and language. It assumes that there is something special and uniquely "literary" in the way literary texts use language. Because of this, the intrinsic attitude concentrates on the language of the text as its central object – considering things like the choice of metaphors, the use of symbols, structure, style, contrasts, images and the development of the plot – to work out what a text means. Although these forms of criticism might sound rather dull and unrewarding, following the intricate paths taken in a text and looking closely at the twists and turns of its language can produce quite remarkable readings and effects. In fact, the very intense scrutiny of the words on the page can result in the most unusual and challenging interpretations of texts, as the multiple and often unclear meanings of each word are weighed and evaluated. As you concentrate on the words themselves, their meanings become not clearer but more ambiguous (or *indeterminate*). This is most obvious when looking at poetry.

For example, there is a sonnet by the English poet William Wordsworth (1770–1850) called "Composed upon Westminster Bridge," which describes all of London, seen from the bridge at dawn, stretched out and radiant: "Earth has not anything to show more fair" and the city "like a garment" wears "the beauty of the morning." The poem finishes with these lines:

> Dear God! The very houses seem asleep
> And all that mighty heart is lying still.

The first meaning of "lying still" is that the city is spread out, not moving, lying motionlessly asleep. But the word "lying" has another meaning, of course: to lie is not to tell the truth. Perhaps the sonnet is implying that the city, *despite* all the beauty of the morning light, is *still* not telling the truth. The sunrise makes London look wonderful, but really the city, "that mighty heart," is still a den of deceit, corruption, falsehood and lies. By concentrating on the language – on the *form* of the text – two separate readings have emerged. On the one hand, London is beautiful, quiet and still in the dawn light. On the other, London *seems* beautiful, but, underneath and despite all this beauty, it is deceitful and corrupt. These readings are contradictory and mutually exclusive: either London is really deeply beautiful and peaceful, or it's actively scheming, lying and dishonest. Which reading you choose depends on the way you interpret "lying still."

All ways of reading share this concentration on language to some extent, but, for the critics who lean toward the intrinsic attitude, studying English is principally a matter of looking at the words on the page with great rigor. This is characteristic, as I argued in chapter 2, of the New Criticism, aiming to find the value of literature in the special sort of manipulation of language that happens, they argue, only in literature. The methods of interpretation that take this intrinsic approach for granted are often still called practical criticism or close reading.

This sort of intrinsic approach to literature is still very influential and important (in fact, some form of the close reading of texts is central to most subjects). When you are asked to do a practical criticism, write an appreciation or appraisal, analyze the main poetic methods, pay close attention to meaning, language and structure, investigate the style or narrative technique or even comment on the author's skill in suggesting unspoken feelings through incident and description, you are being asked to take an intrinsic approach to literature. Even questions on character or plot, although they seem to have a wider focus, usually lead you to take this approach.

Although it might offer some interesting insights, this intrinsic attitude, used alone, does have blind spots and rests upon some rather large assumptions, as outlined in chapter 3. To recap: some critics claim that intrinsic types of criticism lead to "objective" readings, to the ideas that texts can be independent of their historical, social and personal context and that "literary-ness" makes a text a valuable work of art, which is worth studying in its own right. However, even if you claim to be looking only at the text by itself, you bring your own ideas, expectations and experiences to it. How can any judgment of worth be objective?

Extrinsic attitudes: Out of the text

In contrast, extrinsic methods of interpretation take for granted that the literary text is part of the world and rooted in its context. An extrinsic critic considers that the job of criticism is to move from the text outward to some other, not specifically literary object or idea. Such critics use literary texts to explore other ideas about things in the world and in turn to use other ideas to explain the literary text.

Perhaps the most important and widespread sort of extrinsic criticism is the way of reading that puts texts firmly into their historical context. This is why the extrinsic attitude is often referred to as *historicist*. Historicist criticism – and there are many versions of it – uses literary texts to explore or discuss historical issues, and conversely it uses history and context to explain literary texts. In dealing with Shakespeare's *King Lear*, for example, a historicist critic might look through the play to find clues about what was expected of a king at the time Shakespeare was writing, as well as how the ruler and the nation were thought to be woven together. By the same token, a historicist critic might also use evidence from Shakespeare's time and its historical context to explain the play. But historicist criticism is not limited to works from the past: you could use another form of historical criticism to study a contemporary popular novel – a best seller. Looking at the way people behave in the novel, even if it might not be considered a great work of art, would reveal all sorts of interesting contemporary social attitudes. The way two central characters deal with the experience of having cancer and falling in love, for example, might indicate a great deal about attitudes toward the body, health and society, for example.

Many of the newer ways of reading are based on the extrinsic attitude. Critics who use psychoanalysis as a way of reading might understand a literary text as a product of the author's psychology or as a way of understanding parts of the human mind in general. In fact, the work of Sigmund Freud (1856–1939), Jacques Lacan (1901–1981) and other psychoanalysts has been widely used to interpret literature. Those who explicitly champion political positions use literary texts as evidence for wider historical and political arguments. The many forms of feminist criticism use literary texts to explore the roles of women and men, among other things. Other critics start with the text and draw conclusions about, say, nature, humanity or the pitfalls of love. Also extrinsic are approaches that utilize the potential of digitization to scan in seconds more texts than a human could read in order to, for example, identity strings of words or phrases and so make arguments about historical change. Even approaches that consider authors' intentions or their lives display the extrinsic attitude because neither authors nor their biographies are actually *in* the text. The idea of looking beyond a text at "the world" is very attractive to those who emphasize the way in which literature is linked to the world. Many new forms of extrinsic criticism have emerged in the last twenty years or so, as academics have sought ways of reflecting the changes in contemporary society.

Those who oppose extrinsic critical attitudes point to the fact that, in using this approach, you start with a literary text but move away to an object or idea that is *not specifically literary*. They argue that in doing so you do not actually deal with literature itself at all but rather with politics, identity, the mind, history, gender relations, biography and so on. If you approach a text as if it were a piece of evidence for history, opponents say, then it is no different from a treaty, a will or any other kind of historical documentation. If you read a novel to learn about the author, the novel itself is no more than a piece of evidence for a biography and no different from a diary entry. What makes the text special as literature is not of interest.

Contrasting these two attitudes

Looking at the key aspects of these attitudes, as shown in Table 5.1, is a useful way to compare and contrast them.

Table 5.1 **Intrinsic and extrinsic critical attitudes**

Intrinsic attitude	Extrinsic attitude
Into the text	Out from the text to the context
A flat canvas	A window
Literature is worth studying in its own right: it uses language in a unique way.	Literature is worth studying for what it tells us about other things.
"Great texts" are the focus because they have artistic and possibly moral worth.	All sorts of text are worthy of study because they all reveal the world.
Formalism	Historicism
Words on the page	Context
Meanings are often indeterminate.	Context decides meaning.
Practical criticism, "close reading" and New Criticism	Historicism; psychoanalytical criticism; explicitly political criticism; feminism; philosophical criticism; digital criticism; biography and other sorts of criticism
Text stands alone.	Text has meaning only in context.
Knowledge of the text alone	Knowledge of the context (history, author's life and so on)
Style, plot, character	Theme, setting

These oppositions have been the subject of fierce debate, and you will come across signs of this at different levels and in different ways throughout the discipline of English. Both these general attitudes are valid, as are the critical methods they stimulate. Even if they do have blind spots, both have a role to play in English as a whole. Sometimes the most useful works of criticism are produced by a coming together of these two attitudes in different ways.

Thinking about these general patterns helps to orient you by explaining why approaches to literature have developed in the way they have. This introductory guide to critical attitudes also makes it more straightforward for you to draw parallels between different approaches and to explore the presuppositions and blind spots of any particular approach.

Summary

- One way to think about the presuppositions of reading is to divide critical theories into two broad groups or attitudes: intrinsic and extrinsic.
- Intrinsic ways of reading concentrate on *words on the page*. A work is considered separate from the world, and the focus is on its internal features. Critics who support the intrinsic attitude rely on language and structure to decide what a text means.
- Extrinsic ways of reading look beyond the text *to the context*. The literary text is seen as part of the world, and critics move through the words on the page to broader, nonliterary ideas, like history or biography, which are in turn used to explain what a text might mean.
- Both these attitudes have blind spots and gaps. Intrinsic approaches are criticized for assuming that there can be an objective way of reading and for separating literature from the so-called real world. Extrinsic attitudes are criticized for failing to see literature as something special and preferring to discuss nonliterary ideas.
- Thinking about these general patterns helps to orient you when you look at different critical approaches, to draw parallels between different approaches and to explore the presuppositions of any particular approach.

PART II
WHAT WE READ

6

Literature, value and the canon

- Can literature be defined?
- What is literary value?
- What is the canon?
- How does the canon affect you?

If, as I have argued, we think about how we read, we also think about what we read. Debates over the texts you should study and even what literature is have become very important to English studies.

Can literature be defined?

When we go into a bookshop or library or visit an online store, we know basically what to expect in the literature section or category. But if we try to answer the question, "What is literature?" no definition seems satisfactory, as suggested in chapter 4. There are always countless exceptions to every rule. For example, if you define literature as *fiction*, where would you put fact-based writing, such as autobiographies, or plays and novels that portray historical events? Where would you put the poems that claim not to be fictional but to reveal a "higher" truth? If you wanted to suggest that literature *represents the world* (that it was, to use the technical term, *mimetic*), what would you do with the surreal poems, plays and novels that don't seem to represent the world at all? And, after all, don't other forms of writing – historical, scientific – claim to represent the world as well? Literature can't be exclusively something that *tells a story* either. How would this be any different from, for example, a medical textbook telling the story of the symptoms caused by a particular disease or a scientist detailing what happens in an experiment to measure cosmic rays?

You might argue that a work of literature was something that *moved* you or *entertained* you, but what would you call a novel that moved one of your friends but left you cold? You might call it bad literature, but would you say it wasn't literature at all? Again, if you wanted to argue that literature should convey a message, what would you do with writing that didn't seem to convey messages or with literature that was utterly unclear about exactly what message it might be carrying? Besides any of this, couldn't you argue that a song or a sandwich or a bowl of noodles might move you just as much as words on paper?

It is easier to understand literature not as something that can be defined but as something that *overflows* or *escapes* from any attempt to limit it or put it into a box (to define something means to set limits to it). As you try to give it a definite meaning, literature slips through your fingers like water. But then, perhaps literature is not a thing at all, which is why it glides away when you try to categorize it. Reading, after all, is more like a process you are engaged in, something you do. Perhaps literature is more like a verb, a "doing," than it is a noun or thing.

All this is made more complex by the fact that, historically, the category of texts known as literature has changed a great deal. In fact, when the word was first used in the English language, from the late fourteenth century, it didn't mean a type of text at all but rather what we now call literacy, a sort of knowledge of books. By the nineteenth century, the term "literature" did mean a body of writing but included what we would call history, biography, philosophy, sociology, science and much more. It simply meant something written on a certain subject. We still have this sense of literature, but somehow we have invented a separate category called Literature, with a capital "L," which means something quite different.

The philosophical and historical discussions about the identity of literature lead to a fundamental question for anyone studying English: if there is no clear, defined area of study, how do you decide which texts to read? After all, there are far too many books to read in any one lifetime. When we study English, we *choose* our literary texts, or, more accurately and more problematically, *the texts are chosen for us*. Those who have made the choices and shaped the English syllabus have done so with a certain idea in mind: that of literary value.

LITERATURE, VALUE AND THE CANON

What is literary value?

Often when we say "literature," we say it with a capital "L," Literature. Knowingly or not, the term is used to make a *value judgment* about the *worth* of a piece of writing. People say, "This is a truly great novel, it's Literature," or they say, "That's only a thriller [or horror story or romance]; it's not proper Literature." In this sense, Literature doesn't just mean words on pages but rather a certain sort of highly valued and important writing. It used to be taken for granted that people studying literature read only so-called great literature – or Literature. There are even lists of great books that we should read and admire, known as *the canon*. This is why the same novels, poems and plays turn up again and again in literature courses and exams. In no small way, studying English can mean reading, studying and writing about the canon. It affects your courses, exams, results and everything else about English. The content of the canon and the criteria for selection are among the most contentious issues in English – but what is the canon?

What is the canon?

The origins of the canon

Where the idea of the canon came from is unclear: the term itself comes from the Christian Church. Faced with a number of texts about Jesus and the early Christians, with the Hebrew Scriptures, and with disputes about which sources to trust, the Catholic Church decided at the Council of Trent (1545–63) that some of the texts were true sources of "divine revelation" – and so were "canonical" – and that others were not. The aim was to create a list of religious texts that everybody would accept as authentic and authoritative. Eighteenth-century philologists took this desire for "authentic and authoritative" texts into the study of language. Because there were a huge number of forgeries of ancient Greek and Roman texts, these philologists aimed to establish a canon of texts that were really Greek and Roman.

The poets and writers of the Renaissance (roughly 1450–1650) also produced lists, ranking the most important types, or *genres*, of writing (*genre* means "kind" or "type" of literary text). These days, we have many genres of literary text, normally divided not by form but by content. In any bookshop, there are shelves for all sorts of novel genres: thrillers, romances, science fiction, fantasy. These definitions can be even more detailed: a genre of novels set in universities (the so-called campus novel), thrillers where the lead character is a forensic scientist, perhaps. Many types of novel are often dismissed as simply genre fiction and have been excluded from the traditional canon, although there is no reason to suppose that a science fiction novel, for example, would not be as interesting or rewarding to read as a "literary" novel. Each genre has its own generic *conventions*, parts of plot or style that are special to that genre. These occur both in the content (you expect a murder in a whodunit or a marriage at the end of a comic play) and in the style (for example, a spare, terse style in a hard-boiled detective story). Importantly, these conventions also shape our expectations in reading: in "Macbeth Murder Mystery" (1937), a short story by James Thurber (1894–1961), a hotel guest picks up *Macbeth* rather than the crime thriller of the sort she usually reads. But because she is used to reading whodunnits, where, for example, the most obvious suspect is rarely actually the murderer, she discovers that Macbeth is innocent of the murders attributed to him, and the real killer is . . . someone else (you'll have to read it). Occasionally, texts mix up or blur these conventions for effect. In the Renaissance, however, these boundaries and definitions, so important to us today, were just beginning to take shape. The British poet Sir Phillip Sidney (1554–1586) produced a list that classified poetry by type: epic, lyric, comic, satiric, elegiac, amatory, pastoral sonnet, epigram. Epic poetry was the greatest, most enduring and most significant form, while short poems about love were the most transient and insubstantial. By the eighteenth century, it was common to find debates not only over the worth of particular genres of poetry but also over the worth of particular writers. A critic called Joseph Warton (1722–1800) wrote that in "the first class I would place our only three sublime and pathetic poets: Spenser, Shakespeare, Milton" (*pathetic* meant "moving" or "poignant" at this time). Such a reference to "our" poets shows how the idea of literary value was becoming linked to that of nationalism.

The ideas of authority, authenticity, value and nationalism began to come together even more closely in the nineteenth century. Perhaps most influential in the formation of the canon were the many anthologies of poetry popular in the nineteenth century. One of the most famous of these was the *Golden Treasury of English Verse*, compiled by Francis Turner Palgrave (1824–1897), first published in 1861, often reedited and republished. It's still in print and still popular today. But there were many others: Samuel Kettell (1800–1855) produced the first major anthology in 1829, *Specimens of American Poetry*. Henry Longfellow (1807–1882) produced the mammoth thirty-one-volume *Poems of Places* (1876–1879) and Ralph Waldo Emerson (1803–1882) an anthology of poetry called *Parnassus* (1880) (named after the mountain in Greece where the Muses, the demigods who inspired the arts, lived). Edmund Clarence Stedman (1833–1908) produced *An American Anthology, 1787–1900*

(1900), chronologically starting with the "Early Years of the Nation," and Thomas R. Lounsbury (1838–1915), a professor at Yale and a major figure in the origins of English, produced a *Yale Book of American Verse* (1912). The title of the *Golden Treasury of English Verse* is itself very revealing: just as the national treasury has the authority to make financial decisions on behalf of the nation, so a treasury of poetry has taken upon itself the authority to decide which poems should be considered the most valuable by its readers. Just as a nation's treasury contains the material goods – money – most valuable to its people, this treasury contains the poems most valuable to its readers. On the very first page, Palgrave said that he aimed to "include . . . all the best original Lyrical pieces and songs in our language, by writers not living – and none but the best." In judging what to include or exclude, Palgrave used two criteria: the types (genres) of poetry and the "genius" of the poet. No didactic poems (poems intended to instruct), no humorous poems and no narrative poems (those simply telling a story) were allowed in. Only poems relying on what he called "some single thought, feeling or situation" were worthy to be allowed into the *Golden Treasury*. But the poems also had to be "worthy of the writer's genius." This means that writers had to be already recognized as major poets to be included, and the poem had to show off their particular talent. However, is it possible to be, as Palgrave claims he is, without "caprice or particularity" about a writer's talent? In the 1861 edition of the *Golden Treasury*, for example, there were no poems by the radical working-class poet William Blake (1757–1827). Even more significantly, there were no poems by women in the early editions of the anthology. Does this show that poets had to be of a certain class and gender before Palgrave would even consider their poems? Each of the other anthologies I mentioned set criteria, either explicitly or implicitly, for inclusion or exclusion: they decided, for example, what and who counted as "American." As late as 1978, an anthology of American literature stated that by

> American literature we mean literature written in English by people who came to settle in the territory that eventually became the United States of America. We exclude English Canadian literature; the *Relations* written in French by Jesuit missionaries, even those written among the Iroquois on what became United States soil; and the writings of early Spanish missionaries.

Anthology comes from Greek and means "a collection of flowers" (and so of poems), which sounds lovely but, as we have seen, is clearly less innocent than it appears.

WHAT WE READ

The canon and English

These historical threads form the backdrop to the development of the modern canon. What we recognize today as the canon grew up hand in hand with the discipline of English in the early years of the twentieth century. It is here that the assumptions of value, authenticity and authority come clearly into focus and become ever more closely linked with nationalism. Crucial for the development of this was the poet and critic T. S. Eliot (1888–1965). As suggested in chapter 2, although T. S. Eliot is now thought of principally as a poet, his essays of literary criticism in the 1920s were extremely influential; indeed, E. M. W. Tillyard, a critic of the time, described them as "revolutionary." "Tradition and the Individual Talent" (1919) was perhaps *the* most influential single essay in the history of the discipline of English, shaping the subject and making a claim for literature and its study. In the essay, Eliot argues that each artist writes in relation to a tradition,

> not merely with his own generation in his bones, but with a feeling that the whole of the literature of Europe from Homer and within it the whole of the literature of his own country has a simultaneous existence and composes a simultaneous order.

For Eliot, a tradition isn't just the past but a living thing, organized, structured and present in the mind – or even in the bones – of a great writer (always a "he" for Eliot). This "living tradition" of great literature makes up what Eliot later calls an "ideal order," which ranks the great and valuable works. This is clearly a canon. In order to write a great poem, novel or play or to appreciate a great work of literary art fully, Eliot argues that it is necessary that "we" have these works in their "ideal order". If this order is in our bones, it is part of who we are, not something we have to think about. "We" must have internalized and accepted not only the list of works that people like Palgrave decided were great but, more importantly, the criteria that guided their judgment.

Eliot's idea has two consequences. The first concerns what these authoritative texts are authoritatively telling you. An authoritative list of classical texts tells you that certain texts are authentically ancient Greek or Roman and not forgeries or inventions; the authority of books of scripture lies in the fact that they are thought to reveal the authentic word of God. But what authenticity does an authoritative list of works of literature reveal? For Eliot and those influenced by him, what underlies a great literary work, what a great work reveals and so therefore what makes it "authentic" are the values of Western European (and within that, English) culture and life. The canon, he argues, is the "storehouse of Western values." These Western European values are unquestioningly assumed to be *universal human values*, the most important values that apply to all people at all times and in all places.

This leads to the second consequence: if a text doesn't seem to demonstrate these "universal" values or expresses different ones, it is not considered valuable and so is excluded from the canon. Eliot's seemingly innocent metaphor of bones in fact reveals a rather frightening idea. It is as if it is not enough just to study the tradition – it must be in your bones, in your body. If you don't genetically share the idea of the canon and of universal European values underlying it, you can neither properly appreciate nor write great books. In their book *The Decolonization of African Literature*, Chinweizu, Onwuchekwa Jemie and Ihechukwu Madubuike, a trio of African writers and critics, sum this up from their perspective:

> [M]ost of the objections to . . . the African novel sound like admonitions from imperialist mother hens to their wayward or outright rebellious captive chickens. They cluck: "Be Universal! Be Universal!" And what they don't consider universal they denounce as anthropological, atavistic [i.e. reverting to an earlier, primitive state], autobiographical, sociological, journalistic, topical ephemera, as not literary.

They are decrying the idea that what doesn't reveal Western values (masquerading as universal values) simply isn't authentic literature, is not worth reading and couldn't be part of the canon.

How does the canon affect you?

The canon today

The canon is still with us today. It is deeply woven into the fabric not just of English as a subject but into all forms of culture. TV and film adaptations tend to be of "canonical" novels; publishers print "classics"; to count as educated, you are supposed to have read a smattering of canonical novels. Why is the canon such a powerful idea?

First, the canon is a reflection that English always has a social *context* and could never be done in a vacuum. The canon represents the meeting point between (1) judgments of the artistic (or *aesthetic*) value of a text and (2) the presupposition and interests, either implicit or explicit, of those who make those judgments and have the power to enforce them. What makes the issue difficult is that, despite claims to be objective or neutral, it is very, very hard to separate out an artistic or aesthetic judgment from a judgment based on position and interests. These two are interwoven.

Second, the canon is *self-perpetuating*. In English at all levels, the same canonical texts come up again and again, year after year. A person who has studied English and become a teacher often teaches the texts she or he was taught, in part because these texts were presented to them as the most important. As students, you expect to study texts you have heard of and assume are worthwhile, and, of course, resources that support your study, such as websites, IT resources, guidebooks or videos of productions all concentrate on canonical texts, which in a way makes them easier to study. (After all, why would a company produce a guidebook to a novel that only a few people have heard of?)

When a professor is choosing titles for a syllabus, the canon exerts pressures both explicit and financial. Some students will prefer to take a class with texts they have heard of, just as many people prefer to listen to a radio station where they will hear songs by artists they already know. As students know all too well, the cost of books can be a strain. A professor teaching an American literature survey might be weighing a choice between the canonical *The Scarlet Letter* (1850) by Nathaniel Hawthorne (1804–1864) and the noncanonical *Hobomok* (1824) by Lydia Maria Child (1802–1880). Both novels feature female protagonists and seventeenth-century New England settings. Hawthorne's book is widely available in cheap editions, whereas Child's is available in print only in an expensive edition from a university press. Depending on the other texts assigned, that professor might feel pressure to take the less expensive and more canonical path.

Many textbooks for English and books on literature in general assume a familiarity with the canon, which also underlines its centrality. In fact, textbooks from earlier in the twentieth century were often made up literally of lists and descriptions of great books. A more recent version of this is *The Western Canon* from 1994, by the American critic Harold Bloom (b. 1930). This book is a long defense of the idea of the canon and ends with a list of the thousand books (he thinks) every "cultured" person should have read. The canon, then, can be the list of books you expect to study when you do English, and reading the canon can be studying English. The subject and the canon in part define each other.

However, even those who make and publish actual lists of "great books" admit that sometimes the lists can change, as certain books come into and out of favor. But the third reason the canon is so powerful is that it *creates the criteria by which texts are judged*. It sounds like common sense to say that the texts you study must be of high quality and worthy of serious consideration, but these sorts of statements come with no yardstick to measure them; the values that make a work substantial and give it quality are not revealed. New or rediscovered texts are judged by the canon's standards. Saying that a new novel fits the canon because it displays these values reaffirms the idea that an older novel "had" them too. Paradoxically, the canon is not broken up but reaffirmed.

The fourth reason the canon remains powerful is that it is involved with the senses of *identity* to which countries and groups aspire and with the struggle to define identities. As the history of the canon suggested, its development was tied in with the development of ideas about nationality – and of who is properly "American." It is for this reason that Toni Morrison (b. 1931), the Nobel Prize–winning American author, wrote in 1989 that

> Canon building is empire building. Canon defence is national defence. Canon debate, whatever the terrain, nature and range (of criticism, of history, of the history of knowledge, of the definition of language, the universality of aesthetic principles, the sociology of art, the humanist imagination) is the clash of cultures. And *all* the interests are vested.

Because it is canonical texts that are taught, studied, examined, published, sold, bought, performed, made into TV miniseries, adapted for YouTube parodies, updated into contemporary settings in films and for bestsellers and so on, the canon plays a significant role in creating a sense of shared culture and of collective national identity. Deciding which texts are in the canon is all part of deciding who we are and how we want to see ourselves, and a threat to the canon is a threat to national identity. But does the person creating your course ask how you want to see yourself? As Toni Morrison says, all the interests are vested.

Canons tomorrow?

Because there are simply too many books to read within the limits of any course, decisions have to be made. Every course has to have a curriculum, which creates, straightway, a sort of a canon. However, since the canon and the texts you study are so important, these decisions stimulate furious, heated and often public debate. However, the idea that there is one canon is perhaps breaking down, as new approaches and new ideas affect which books people decide to teach and study.

Nevertheless, the *idea* of the canon is still very powerful. If the single monolithic canon is breaking down, an array of separate canons has taken its place. There are canons of African-American writing, of women's writing and of science fiction, for example. Texts that were previously marginalized by the canon now are considered important and have canons of their own. Those studying and teaching English have more freedom to choose one canon over another. More than this, each canon, each literary tradition, has developed, within the broader questions that all literary writing begs, its own questions and approaches. In a discussion about African-America literature, for example, one of the leading African-American critics, Henry Louis Gates, Jr. (b. 1950), wrote that

> a funny thing happened on the way to the Civil Rights movement. Black writers started reading and revising each other's work, situating their representations of their own experience and those of other black people, in the tropes and metaphors of other black writers. That is what a literary tradition is: it is a body of texts defined by signifying relations of revision.

And Lovalerie King and Shirley Moody-Turner identify some of the questions that this African-American tradition, or canon, poses especially acutely in a contemporary setting:

> [Issues of] race and authenticity, the literary versus the popular, and ideations of "post-racial"... "post-Civil Rights"... When did blackness begin? What is authentic blackness? Is Blackness over?... What are the pros and cons of the commercialisation of African American literature?... How do we teach contemporary texts as part of a 250 year old tradition?

These sorts of issues are involved in both forming and coming to understand a canon or tradition of literary texts. The criticism shapes the canon, and the canon shapes the criticism. Perhaps we can begin to think of a canon literary texts and the criticism that goes with them as sort of conversation over time, focusing on certain themes and ideas as they develop. But we still can think about who sets the terms of that conversation.

The power of the canon makes it essential for us to question what we read. How did it get into the canon? Why? What were the values of those who chose the text? As part of this process of questioning the canon, I will now turn to the figure at the center of the canon and (some might argue) at the center of the discipline of English itself: William Shakespeare.

Summary

- No definitions of literature seem to be adequate: literature overflows or eludes any attempt to categorize it.
- Often unknowingly, we make value judgments about writing: literature comes to mean a certain sort of highly valued and important writing.
- The list of great books that we should read and admire is known as the canon. The process by which texts are chosen to be part of the canon depends upon (questionable) ideas of authenticity, authority, nationalism and literary value.
- The canon is still with us today, woven into the fabric of Western culture. It is the meeting point between artistic judgment and wider presuppositions; it is self-perpetuating; it sets up the criteria by which texts are judged; it is involved with our sense of identity.
- The canon appears to be changing and developing into canons; however, it is still vital to know how and why any canon is constructed.

7

Castle Shakespeare

- Why is Shakespeare so central to studying English literature?
- What are the traditional arguments for studying Shakespeare?
- What are some of the new ideas about studying Shakespeare?
- How do these ideas affect the way we study Shakespeare?

Chapter 6 examined the canon in general, and this chapter is going to examine debates about the texts that have been assumed to be the very center of the canon – the plays of William Shakespeare (1564–1616). Debate rages over approaches to Shakespeare, but this discussion is rarely explained to students.

Castle Shakespeare

Shakespeare has become a literary institution, seen by many as the unquestionable center of English studies (indeed, one college president joked that he would rename his Department of English as the Department of Shakespeare). In her book, written for her niece, *Letters to Alice*, the novelist Fay Weldon (b. 1931) suggests that writers "build Houses of Imagination" in "the City of Invention." This city has an "all male suburb of sci-fi," a "Romance alley" and "public buildings and worthy monuments, which some find boring and others magnificent." The city is a particularly interesting metaphor for literary value, since, just as in any city, some districts are different: some are considered "better" than others, for example. Weldon writes that at the "heart of the city is the great Castle Shakespeare. You see it whichever way you look. It rears its head into the clouds reaching into the celestial sky, dominating everything around." Although the huge castle is a "rather uneven building, frankly . . . shoddy, and rather carelessly constructed in parts," Weldon writes that it "keeps standing through the centuries and, build as others may, they can never quite achieve the same grandeur; and the visitors keep flocking, and the guides keep training and re-training, finding yet new ways of explaining the old building." Weldon is showing us the way Shakespeare holds his place at the heart of the canon, while apparently other authors try in vain to achieve his stature and literary critics offer new ways of approaching his work. But a castle is not a peaceful place: it is designed to withstand sieges, to play a role in war. Medieval rulers built castles as a sign of ownership and authority, and they aimed to frighten their subjects into submission. This is true with Shakespeare's work too: it is constantly controversial. The institution of Shakespeare divides as much as it unifies, and this is why Weldon's image of Shakespeare as a castle is so apt: a castle means security for those living within but is imposing and even threatening to those outside.

But the institution of Shakespeare stretches well beyond the world of literature. It is an American institution: the plays are the most widely and frequently performed across the nation; they are frequently made into movies or adapted (the charming *10 Things I Hate About You* (1999) moves Shakespeare's *The Taming of the Shrew* from Padua in Italy to Padua High School); the world's largest collection of Shakespearian material, the Folger Shakespeare Library is on Capitol Hill close to the Supreme Court and Library of Congress in Washington, D.C.; Shakespeare's works are used, reused and parodied in literature, on TV and online and constantly alluded to; his words are adapted daily in speeches, headlines, advertisements or titles (the title *The Fault in Our Stars* comes from *Julius Caesar* but also references *Romeo and Juliet* as unfortunate "star-cross'd lovers"). Shakespeare's phrases have even entered the language: as the journalist Bernard Levin pointed out, if you have ever not slept a wink, refused to budge an inch, made a virtue of necessity, knitted your brows, stood on ceremony or had short shrift, cold comfort or too much of a good thing, you're quoting Shakespeare.

And, of course, his plays are required reading in high school classes and probably the most studied subject on university and college English syllabuses – which makes him by far the most studied single author.

However, while it's certainly traditional, like many things in studying English it's not immediately obvious to everyone *why* you should have to study Shakespeare. President Bill Clinton wrote that when he first had to study Shakespeare, "I was not overjoyed . . . to make his acquaintance." A British educationalist, John Yandell, asked a group of 12- and 13-year-olds why they would be studying Shakespeare in the year ahead. They gave various answers: "It's part of our education"; "Because he was the best"; "You don't hear of no other people who do plays like him"; "When his plays came out, the first people who saw it thought it was really good, but it's hard for us to understand it because times have changed"; "We've got to because of the exam; because the play is written in English." These different answers are all, in fact, quite similar. To say that you have to study Shakespeare's plays for the test or because they are on the curriculum or simply because they're in English is only to say, really, that you study Shakespeare's plays because you're told to. The students who say, before they've actually studied Shakespeare, that he is the best or that the first people who saw his plays thought them excellent also sound as if really they're answering "because we're told to": they have been *told* that the plays are the best or were much appreciated by early audiences, so they have taken Shakespeare's excellence for granted. John Yandell interviewed teachers, too. as one responded,

> when kids go "I hate Shakespeare" I can honestly say "I really understand that, I'm not telling you that it's brilliant." And sometimes they ask "Why have we got to study this?" and the personal side of me thinks "I haven't got an answer for that – I had to, you have to" . . . it's never very satisfactory.

Several other teachers felt the same – "I had to, you have to": it's just a convention. The same question arises: why? There must be better reasons to study Shakespeare than because you have to. Certainly many critics and academics have tried to offer reasons. As with many other issues in English, the study of Shakespeare is the focus of a highly contentious debate, which has not yet filtered down to most students. This debate has been running since the mid-1980s, when all that was "traditional English" began to come into question. As Shakespeare was (and still is) seen by so many as central to English courses, the debate over why he should be studied has led to some particularly fierce arguments. Roughly speaking, there are two camps: on the one hand are those who might be called the *traditionalists*; on the other are a number of critics whom Jonathan Bate describes as the *New Iconoclasts* (an iconoclast is literally an "icon breaker" and means a person who attacks established ideas). As you might expect, there is no neutral view on this: both camps have presuppositions that determine their opinions. Importantly, this very basic question – why do we have to study Shakespeare? – has parallels with many of the other controversies about Shakespeare, so it is central to understanding these too. The rest of this chapter sketches their arguments and then outlines what effect these have for studying English.

Studying Shakespeare: The traditionalists' argument

Shakespeare's friend Ben Jonson (1572–1637) wrote that Shakespeare is "not of an age, but for all time": this might be the motto of the traditionalists' argument for the study of Shakespeare. Simply, they argue or assume that Shakespeare's plays are the greatest of literary texts, which makes the study of them invaluable. It is possible to break this argument down into three parts:

- The artistic (or aesthetic) worth of Shakespeare's plays
- The values taught by Shakespeare's plays
- The universal appeal of Shakespeare's work

The traditionalists' argument suggests that Shakespeare's plays are unarguably the pinnacle of literary art and that their aesthetic worth cannot be rivaled. The National Endowment for the Arts, for example, runs a huge "Shakespeare in American Communities" program, reaching out not only to towns, schools and theatres but also to military bases and prisons. A student guide *Studying Shakespeare*, asks "Why study Shakespeare?," then answers by saying "We need look no further than the opening exchange of *Hamlet*." The authors offer a critical analysis of the passage and repeat the process with passages from the plays *As You Like It* and *King Lear*. This is as if to say, "If we just look at a passage of Shakespeare, its brilliance will convince us that Shakespeare is the best and so deserves more study than the work of other writers."

Traditionalists also argue that Shakespeare is the best teacher of values. In chapter 2, I discussed poet and critic Ralph Waldo Emerson's views on literature as a way of teaching people how to live and how to be a community: the critic Michael Bristol argues that, for Emerson, Shakespeare was a founder of "a specifically American experience of individuality and of collective life." Emerson argued that, rather than being "Americanised" when read in America, such was Shakespeare's genius he "Shakespearianised" America. In his book *Representative Men* (1850), Emerson wrote of Shakespeare:

What point of morals, of manners, of economy, of philosophy, of religion, of taste, of the conduct of life, has he not settled? What mystery has he not signified his knowledge of? What office, or function, or district of man's work, has he not remembered? What king has he not taught state . . . ? What maiden has not found him finer than her delicacy? What lover has he not outloved? What sage has he not outseen? What gentleman has he not instructed in the rudeness of his behavior?

Shakespeare is seen as a font of wisdom and a source of truth about human behavior, good and bad. For traditionalists, literature teaches values and ideals, and Shakespeare's works are the highest form of literature. This means that to study Shakespeare is not just to study one man's work but to study the human spirit at its finest.

What is particularly interesting is that people with very different values find their own values reflected in Shakespeare. For example, in his book *Shakespeare*, the critic Kiernan Ryan describes how the plays "sharpen our need to forge a world from which division has been purged." For him, Shakespeare's plays are radical, suggesting that the established order needs to be shaken up and reformed. In contrast, the American Council of Trustees and Alumni, a group that monitors higher educational curricula for the presence of certain traditional fields of study, argues that Shakespeare offers education in "leadership" and the unity of the nation: Shakespeare provides "a common frame of reference that helps unite us into a single community of discourse." For the businesspeople Norman Augustine and Ken Adelman, "the Bard's profound insights into human nature" mean that he is a useful business guru:

> [Unlike] most contemporary plays or other artistic creations, *The Merchant of Venice* extols business and shows respect for corporate executives and admiration for commerce in general. Within its story are sharp examples needed by every businessperson who has asked, "When should I take a risk – and how best can I manage it?"

These three examples focus on the "universal" values the plays are said to present.

This leads to the final part of the traditionalists' views: that because everybody is moved and affected by Shakespeare's plays, Shakespeare embodies universal values and has something to say to all people at all times and in all places. Traditionalists often suggest that anybody seeing or reading the plays feels that Shakespeare is speaking to them and their innermost thoughts.

In a lecture in 1985, the American poet, writer and activist Maya Angelou (1928–2015) described her love for Shakespeare. Growing up in poverty in the southern United States and experiencing American racism, she said that she felt Shakespeare spoke to her so completely that she knew "William Shakespeare was a black woman." The traditionalists argue that Shakespeare's works should be studied precisely because of this universal quality. They might be said to express the basic emotions, thoughts, ideas, hopes and fears of everybody in the world.

For the traditionalists, Shakespeare's plays are like a stars – beautiful, remote, independent of the earth and worldly concerns, to be wondered at and admired. Yet, like medieval sailors navigating by the night sky, we are given direction by the stars. They give us core values, and by studying Shakespeare we learn those values.

Using Shakespeare: The cultural materialists' argument

Opposed to the traditionalist arguments are critics and thinkers who are sometimes described as *cultural materialists*. A cultural materialist critic is principally interested in the way material factors – like economic conditions and political struggles of all sorts – have influenced or even created a text. In turn, they argue that any text can tell us about these material conditions. Because their interest is in the context of works, they argue that all works of culture – here, Shakespeare's plays – are involved with politics and the world. (This reveals the extrinsic attitude discussed in chapter 5, where critics look beyond the text to other nonliterary ideas.) For a cultural materialist, Shakespeare – both the plays and the institution – is a construct of present-day political, cultural and economic interests, rather than a transcendent spring of beauty, wisdom and values. Where traditionalists understand Shakespeare as a beautiful remote star, cultural materialists see his plays as trees, growing from the soil of political concerns in the world. They reject the "traditional" claims made for Shakespeare's plays.

Is Shakespeare "simply the best"?

To begin with, cultural materialists oppose the aesthetic worth argument and deny that Shakespeare is "simply the best." In addition to suggesting that "the best" in literature is not as straightforward as it seems – Whose best? Who decided? Why? – cultural materialists have two arguments. First, they describe the development of Shakespeare's reputation, showing that the idea of Shakespeare as the best is not the result of the quality "shining through" but instead the result of historical events. Second, they compare Shakespeare's reputation with the reputation of other writers to highlight the elements of historical chance.

The story of how Shakespeare the Playwright became Shakespeare the Institution is a long one, and a number of easily available sources cover it in detail (see Further Reading). Roughly, it suggests that, although Shakespeare was successful during his career as a dramatist, he was not seen as outstanding. For example, Shakespeare was buried quietly in 1616: in contrast, when his friend and rival Ben Jonson died in 1637, a crowd followed the coffin to St. Paul's Cathedral. Historians of Shakespeare's reputation argue that its first boost came in 1660. From 1642 to 1660, during the Civil War and Commonwealth, theaters first in London, then throughout England, were closed because the country's rulers – Oliver Cromwell (1599–1658) and Parliament – considered plays immoral. In 1660, the theaters were reopened. Lacking any recent material, theater owners and managers were forced to revive plays from the past, including Shakespeare. A handful of editions of Shakespeare's plays were brought out by theater managers for use in the theater. However, as Gary Taylor points out in *Reinventing Shakespeare*, a very readable study of Shakespeare's changing reputation, between 1660 and 1700 as many as thirty editions of plays by Shakespeare's near contemporaries Beaumont and Fletcher were published. This shows that Shakespeare was not seen as the most important playwright. Nevertheless, toward the end of the sixteenth century and beginning of the seventeenth, Shakespeare's reputation began to grow. As the market for books grew, so did editions of Shakespeare: there were editions in 1709, 1725, 1733, 1747, 1765 and 1768. In fact, it became quite the thing for somebody with literary ambitions to edit Shakespeare as a marker of their own importance and seriousness.

By the beginning of the nineteenth century, the growth of the Romantic Movement in the arts helped to foster Shakespeare's reputation. Romantics considered the "creative force" to be vitally important, and they saw Shakespeare as a leading example of creativity. His work was read more widely, and the characters of his plays began to take on their own life. Shakespeare was more popular in America than he was in Britain. John Quincy Adams (1767–1848) wrote that "at ten years of age," he was "familiarly acquainted" with Shakespeare's "lovers and his clowns," and, after abandoning other literature, he "continued to read the Bible and Shakespeare." Alexis de Tocqueville, in his famous travels around the country during the 1830s, found

Shakespeare everywhere: there's "hardly a pioneer's hut that does not contain a few odd volumes of Shakespeare. I remember that I read the feudal drama of *Henry V* for the first time in a log cabin." Throughout the nineteenth and early twentieth centuries, Shakespeare was a common point of reference for American society. Indeed, in the famous *Readers* edited by William Holmes McGuffey, which began in 1836 and dominated American literary education, Shakespeare was the dominant and most frequently excerpted writer. While Shakespeare's importance has slightly decreased since the Second World War, his work is still a dominant force in cultural life.

Shakespeare's reputation has been caught up in a snowball effect. As "everyone" seems to agree that Shakespeare has the highest prestige, people try to associate themselves with the institution of Shakespeare as a sign of their own value. For example, if aspiring theater directors want to show that they can be considered highly talented, they take on the "hardest" challenge of the "greatest" plays: Shakespeare. Actors often say they knew they had made it when they played their first Shakespeare role. Films and TV shows use Shakespeare to sound serious, and series like *Breaking Bad* or *The Wire* are compared to Shakespeare's work. Film studios make Shakespeare-like films to prove their artistic credentials or adapt them to show their relevance. And if such people keep demonstrating that they see Shakespeare as the best, others will keep believing it.

However, looking more closely at this history of Shakespeare's reputation, the cultural materialists argue that the assumption that Shakespeare is the best relies not simply on the quality of his work but on historical chance. This is highlighted by comparing his work to that of other writers. A number of authors could be considered just as great as Shakespeare, but, lacking the support of an empire and all the cultural power of England and the English over four hundred years, they simply don't have the same reputation. The Athenian playwright Sophocles (c. 496–c. 406 BCE) had a major influence on the genre of tragedy, but only seven of the 120 or so of his plays survive. The prolific Spanish writer Lope de Vega (1562–1635) was a contemporary of Shakespeare. He wrote many more plays than Shakespeare did, for a similar audience, and they were very popular. Jonathan Bate takes up this case in *The Genius of Shakespeare*, pointing out that "Spain went into decline and Lope was not translated. The whole of Shakespeare has been translated into a score of languages; less than ten per cent of Lope de Vega's surviving plays has ever been translated into English." According to Bate, the decline of Spain as a political power led to the failure of Lope de Vega to survive as a "great world writer."

The philosopher Ludwig Wittgenstein (1889–1951) has an interesting view on this:

> [While] I hear expressions of admiration for Shakespeare made by the distinguished men of several centuries, I can never rid myself of a suspicion that praising him has been a matter of convention, even though I have to tell myself that this is not the case.

He goes on to say that "an enormous amount of praise has been and still is lavished on Shakespeare without understanding and for specious reasons by a thousand professors of literature." Later, he notes that

> I am *deeply* suspicious of most of Shakespeare's admirers. I think the trouble is that, in western culture, he stands alone and so, one can only place him by placing him wrongly.

That is, it is because Shakespeare is taken to be great that his work can't be properly located, defined or pinned down. Shakespeare is great, perhaps, despite his admirers, rather than because of them.

Does Shakespeare teach values?

The second traditionalist claim I discussed was that texts transmit universal values applicable to all people at all times ("not for an age, but for all time"). The cultural materialists oppose this, saying that the time and place in which works were written and are being read are vitally important. A great work isn't "neutrally" great but has been acclaimed as great for certain reasons. A cultural materialist might ask, suspiciously, why any particular judgment was made at any particular time or why that play was popular at that historical moment. During the Second World War, for example, Maurice Evans (1901–1989) directed a version of *Hamlet* that aimed to show Hamlet as a character "in whom every G. I. would see himself reflected – a man compelled to champion his conception of right in a world threatened by a domination of evil." "Action," Evans wrote, "would be the keynote of our production," and each soldier was

> in his own way a Hamlet, bewildered by the uninvited circumstances in which he found himself and groping for the moral justification and the physical courage demanded of him. If we could succeed in making the parallel of Hamlet's perplexities apparent, the significance of the play to our audience would be magnified.

This version, which came to be known as the *G.I. Hamlet*, was a huge success, played on Broadway and toured across the country. Whereas a traditionalist might argue that Shakespeare speaks to everyone, a cultural materialist – or a theater director, perhaps – argues that context, class, ethnicity, gender, age, education and so on make a great deal of difference. No text can speak in the same way to everybody: some people might even say the text doesn't speak to them at all.

For a cultural materialist, it is no surprise that people on both the right and the left can find their values reflected in Shakespeare. They argue that there is no one "right" meaning in Shakespeare: we each read into the plays what we will, depending on our worldviews. What is interesting to the cultural materialists, if there is no essential meaning or universal value to be sought, is the way in which Shakespeare's plays are *used*: plays can be used to transmit views, as well as to reflect them. In his very accessible and witty books *That Shakespeherian Rag* and *Meaning by Shakespeare*, Terence Hawkes (1932–2014), a leading figure in this movement, argued that there is no "real" Shakespeare and that his plays are not "the repository, guarantee and chief distributor next to God of unchanging truths." "Shakespeare" is only the name for a cultural tool to convince people of a series of ideas. As an institution, Shakespeare has a great deal of authority: if someone wishes to persuade you of an idea, calling on Shakespeare as evidence seems to give that idea more strength.

Even more interesting is Hawkes' idea that the institutionalization of Shakespeare turns the plays into ciphers. In *Reinventing Shakespeare*, Gary Taylor compares Shakespeare to a black hole:

> Shakespeare himself no longer transmits visible light: his stellar energies have been trapped within the gravity well of this own reputation. We find in Shakespeare only what we bring to him or what others have left behind; he gives us back our own values.

For Taylor, all the work done on Shakespeare by academics, teachers, critics, students, theater directors, actors, filmmakers and so on has obliterated Shakespeare, and what is left is merely a reflection of their own values. Sometimes it seems that Shakespeare is so much part of our society that we don't even need to read his plays: you can see a film of *Romeo and Juliet*, and it will give you an idea of what it's about. You may feel you know the play, but in fact you have seen someone's interpretation of the text, with issues emphasized by the director because those were important to her or him. If this is the case, you are learning more about the director's values than you are about

Shakespeare's play. And if you then read the original text, it may well be harder to interpret it another way, once you have certain ideas – presuppositions – in your mind. There is so much talk about Shakespeare and so many ideas about the plays crop up in everyday life, that it is perhaps impossible to think about the text itself rather than what people have said about it.

Another case of Shakespeare reflecting values is the link made between class, education and Shakespeare. For example, David Hornbrook writes that, for most people, Shakespeare "is inescapably associated with social snobbery." Students (especially in school) who enjoy Shakespeare are usually the "academic" ones, the "literary A stream." As this is usually a minority of students, Shakespeare is thus seen as elitist. The central role of Shakespeare in the examination system and its links with success and rewards in education lead to an understanding that Shakespeare divides the good from the bad. Knowing about Shakespeare is a badge of admission into a certain group. Admiring Shakespeare creates a "we," a sense of shared identity, and to dislike Shakespeare is seen almost as a declaration that you are not one of us and not, for example, patriotic. You might also notice that lots of guides to Shakespeare use "we" throughout – "through studying Shakespeare we learn," for example, or "we need look no further." This seems innocent enough, but any "we" ("us over here") needs a "they" ("them over there") in order to define itself: Shakespeare is used as a key tool of that definition. It may be wise to wonder about who this "we" – teachers, students, academics, corporations, the government – actually is and what other ideas this "we" might be passing on to you. This is not to say that the "we" always has to be elitist. Indeed, in *The Genius of Shakespeare*, Jonathan Bate argues that Shakespeare has been used as a subversive antielitist force. There, for example, is a 1968 version of *The Tempest* by the Martinique-born writer Aimé Césaire (1913–2008), in which the play is retold from the point of view of the slave Caliban. The "wise old man," Prospero, is seen as a totalitarian slave owner. Shakespeare here is being used to oppose racism and highlight a slave-owning past.

Does Shakespeare have a universal appeal?

Cultural materialists also question the traditionalists' third supposition that Shakespeare has universal appeal. Does everybody even understand Shakespeare the first time they read him, let alone have a strong response? There are, as might be expected, formidable resources for helping to teach Shakespeare's plays at college or secondary school: notes, websites, support from the National Education Association and so on. The paradox is, of course, that if Shakespeare did speak to everybody, all these efforts to make his work seem accessible and exciting simply wouldn't be necessary in order to make his work seem universal. This is not to say that everything you study should come easily – it shouldn't – but that it's not always so clear whether it is made to seem universal or it actually might be universal.

For the cultural materialists, then, it is impossible to get to a "real" Shakespeare. Moreover, Shakespeare the institution is never innocent or neutral. More than any other name – more than any other series of literary texts – Shakespeare is *used*.

The effects of this debate on studying Shakespeare

These academic arguments about Shakespeare's reputation and the way in which the plays are understood have direct effects on the way you study Shakespeare. A more traditionalist view suggests that you might simply look at plot, character and themes (as any study guide will show). The plot is studied because it is the easiest to understand. The characters are studied because it is assumed that Shakespeare still "speaks" to us through the characters. And the themes are studied not just because studying English has traditionally concentrated on finding the "message" in a text but also because the themes of Shakespeare are "universal" and so reveal "universal values."

However, the cultural materialist viewpoint brings with it a whole range of fascinating new questions you could use to approach Shakespeare. Some of these questions might focus on how Shakespeare's plays are used. Why do productions of his plays differ? What lies behind the differences in film versions of the plays? Others might explore the cultural power of Shakespeare. Why are quotations from Shakespeare found throughout the national press? Why do so many novels, from all genres, use Shakespearean quotations as titles? Other questions might focus on the editions themselves. Should editors modernize the spelling of the plays or leave it in the original? What is at stake in this choice? Why do teachers tell you to read one edition rather than another?

In relation to the plays themselves, there is an even wider range of questions. In a book for teachers by Susan Leach called *Shakespeare in the Classroom*, the author suggests the following examples:

- Who holds the power in the play?
- What is the economic basis of the play?
- Is the power held/obeyed/challenged/overthrown?

- What is the framework within which the plays operate?
- Is it possible to make easy judgments about the behaviors of any character?
- How does gender work in the play?
- How are women presented?

These questions, which don't take the greatness of Shakespeare or the universal values of his plays for granted, move a long way from the familiar trinity of plot/character/themes.

Traditionalists and iconoclasts in other debates

I suggested that the very basic question of "Why study Shakespeare?" paralleled other controversies. Indeed, these sorts of debates have led to wider discussions about adaptations, for example. Ayanna Thompson's book *Passing Strange* analyzes the issue of Shakespeare and race in contemporary America. She identifies a tension between more traditional views of Shakespeare that seek to "reform and unify" the population and more iconoclastic ones that seek to use Shakespeare to highlight precisely what is disunified and fissiparous in society. For Thompson, there is no "raw" or "real" Shakespeare – all versions of his plays are just that: versions. While much of the rhetoric about teaching and performing Shakespeare in schools and prisons, for example, praises the "authenticity and authority" of Shakespeare as universal, Thompson does not find these qualities in the productions that reflect a fantasy about Elizabethan and Jacobean England or ideas about timeless values: instead, she finds them in companies, like Will Power to Youth in Los Angeles, that reflect local and specific issues – particularly regarding matters of race. This sort of argument, too, exists over the ways in which, more generally, people think about "global" adaptations of Shakespeare. For some people, the plays exist to be cherished, and productions of them must have some degree of fidelity to the idea of the text: for others, the plays are just the starting point. They exist to be performed, acted, changed, as one might change a recipe or improvise a piece of music around a tune. Any performance of Shakespeare is an adaptation of the play text, and some – for example, comic Hollywood films or productions set in China – go much further in improvising or challenging the text than others; indeed, you might well ask when a Shakespeare play stops being a Shakespeare play. (Of course, there is no right answer to these questions, but exploring them tells us more about Shakespeare, about ourselves and about others).

Another version of this debate appears in a current controversy between two different groups of scholars. On the one hand are new historicists, who are keen to put Shakespeare's plays into their historical context in order to understand them as works from the past. "Dramatists are best understood in relation to their time," writes James Shapiro, author of *1599*, a detailed study of one momentous year of Shakespeare's life. Perhaps the most famous current exponent of this is the critic Stephen Greenblatt, who invented the term "new historicism." He not only writes beautifully but also consistently places Shakespeare's works into the complexities of his time. For these people, Shakespeare's greatness is interwoven into his context. On the other hand are those who call themselves presentists: presentism is used by historians to condemn people of seeing things only from the present perspective, without the correct historical context. In some ways, this is a foolish accusation because we can see things from the past only from our current perspective (so all forms of history or context setting are somehow presentist). So in contrast, rather than thinking of the present as an obstacle to be avoided in our understanding of Shakespeare, presentists think of it as a factor actively to be sought out and something to which we should pay close attention. This sort of approach not only recognizes the artistic power of Shakespeare's plays but also, as Ewan Fernie argues, that the plays require us to be "responsible," that is, to respond to them. In part, this can mean that, for example, the anti-Semitism of *The Merchant of Venice* is not simply a historical curiosity but has to be faced; in part, it can mean that questions about power and politics posed in *Julius Caesar* still have to be thought through.

Our views on Shakespeare are still changing in other ways, too, even four hundred years after his death. For example, since the 1970s it has been usual to think about Shakespeare plays as works of literature to be performed (they are *plays*, after all, meant to be seen and heard rather than read like novels). However, many of the plays are in fact too long to be easily performed. Recently, one scholar, Lucas Erne, using historical sources, suggested that the

> chief reasons why Shakespeare wrote excessively long plays from the point of view of the public stage is that his economically secure position as a shareholder in his company allowed him to do so and that he cared for a readership of his playbooks [the texts of his plays] and – given his success as print author – knew there would be one.

That is, Shakespeare was like a filmmaker who edits one version of a film down for the movie theater and produces another, much longer one, for the lucrative DVD release, with "restored" scenes, special features and so on. If the versions of the plays we have, then, were meant to be read as much as performed, this changes our view of Shakespeare's works (and caused a riot among Shakespeare experts). Scholars also now know that Shakespeare, like most playwrights of his period, was much more of a collaborative writer than previously thought. And even the number of his plays has changed: in 2011, a version of one of Shakespeare's lost plays, *Cardenio*, written in 1613 with Fletcher, was performed. Emma Smith explains that the source material is known [*Don Quixote*, the great Spanish novel by Miguel de Cervantes (1547–1616)], that there was an adaption in 1727 (called *Double Falsehood*) and that some of the music of the play has recently been identified; this allowed for a "reconstruction" – to some degree – of the text of the play.

Exploring these debates over Shakespeare shows that thinking about what we read, like thinking about how we read, leads to all sorts of questions about how we see the world. Asking "Why study Shakespeare?" leads directly to questions about the relationship between art and politics or between literature and history, and it is interwoven with important issues like gender, sexuality, class, ethnicity and national identity. Despite being opposed to the traditionalist view, the cultural materialist approach doesn't necessarily argue that Shakespeare isn't worth studying or that all artistic values are relative, but it does insist that it's worth questioning assumptions about the author and the plays. However, it is vital not just to assume Shakespeare's greatness but also to think about how we construct it: not just the plays but *how we look* at the plays. Castle Shakespeare may be full of tourists, but it is still a site of conflict.

Summary

- Shakespeare has become an institution, not only in literature but also in cultural life. It's almost impossible to avoid the institution of Shakespeare.
- Traditionalists argue that Shakespeare should be studied because of the aesthetic worth of his work, because he communicates values shared by everyone and because he has universal appeal.
- Cultural materialists are more interested in the way the institution of Shakespeare is related to politics and history. They argue that he is considered the best through historical chance, that the values we see in Shakespeare depend upon our own ideas or those of others who "use" the institution, and that the plays do not speak to everyone. Cultural materialists argue that Shakespeare is the name for a key cultural tool used to convince people of a series of ideas.
- Whichever approach you agree with, the debate shows the importance of thinking about how you look at Shakespeare's work.

PART III
READING, WRITING AND MEANING

8

The author is dead?

- Who decides what a text means: the author or the reader?
- What is the traditional view of the author, meaning and the text?
- What are the problems with this view?
- How else can we determine the meaning of the text?
- Why has the author always seemed so important?
- What are the consequences of all of this?

Having looked at how we read and what we read, I'm going to move on to other debates in English that center on questions of literature, meaning and how we see the world. This chapter is about the relationship between texts and meaning, between authors and readers.

How important is the author in deciding what a work of literature means?

At first, this might look like a silly question: after all, the writer *wrote* the text and must have meant something by it. However, for literary critics this question has been the focus of one of the most heated debates of the last seventy years. Very roughly, the debate has two sides: those who believe that *authorial intention* – or what the author "meant" – is central to working out the meaning of a text and those who believe that the author's intention does not determine what a text means and that any understanding depends on the individual *reader's interpretation*. Perhaps the most influential figures on this second side of the debate were the critic William Wimsatt (1907–1975) and the philosopher Monroe Beardsley (1915–1985), who wrote an extraordinarily influential article called "The Intentional Fallacy" in the mid-1940s at the height of New Criticism. The argument was given a further (and more melodramatic) twist by the French writer and critic Roland Barthes (1915–1980), who wrote a piece announcing "The Death of the Author." While the whole discussion is more formally known as the debate over the intentional fallacy or over authorial intention, it is often referred to as the author is dead debate, in an echo of Barthes's title (another example of the sort of phrases discussed in chapter 4).

For authorial intention: The authority of the author

It seems only common sense that a literary text means what the author intended it to mean. In everyday speech, when you say, "The dining hall is on the left," you mean, after all, that the dining hall is on the left. But literary texts are not ordinary speech (even if, at first sight, they look like ordinary speech), and common sense is often the pretext for taking an idea for granted. If the aim of studying literature is to think about *how* we read, then it is exactly this sort of presupposition that needs to be examined. What, then, are the ideas wrapped up in this so-called common-sense attitude?

Those who share this attitude believe that the text means what the author intended it to mean and nothing else. The text itself, they imply, is like a code, in which the author has encrypted her or his meaning. In reading, the reader decodes the language of the text to find the ideas that the writer has hidden within. A diagram to express this might look like Figure 8.1.

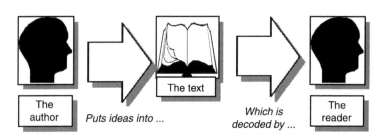

Figure 8.1 **The traditional approach**

This seemingly simple idea – that reading a poem or a novel, seeing a play, is just decoding what the author intended – implies at least four presuppositions that have profound consequences for the study of English.

1. Meaning

If a text is understood as the encoding of the author's intention, it leads to the assumption that the text has one definite meaning, just as a code has a definite meaning. Once having cracked the code, the reader has explained the text and solved the riddle: the reader can give a final and accurate account of meaning, and there is nothing more to say. However, works of literature often have ambiguous phrasing and seem to offer two (or many more!) meanings. Then people who argue this point of view suggest that the author intended to be ambiguous and meant both things at once (with the implication that she or he was very clever to be able to do that). In general, this assumption leads to essay and exam questions like, "How does Shakespeare convey the strengths and weaknesses of Othello's character?" If the reader sees Othello as both courageous and credulous, it is because Shakespeare intended it to be so. The assumption also leads to some interpretations of texts being described as wrong because they are not considered to be what the author intended.

2. Biographical evidence

If you accept that what the author intended is what the text means, it seems possible that you could understand a text without even reading it. Imagine finding some evidence – a letter from the author to a friend, for example – that says, "I mean my novel to be about the conflict between good and evil." Then you could say, "This novel is about good and evil. I know this because the author said so!" It would be like seeing the original message before it was put into code. This sort of interpretation, autobiographical criticism, uses the writer's life story, through letters, diaries and so on, to explain the text.

3. Authorial presence

All these assumptions rely on the idea that the author is, in some strange way, present in the text, actually there. Through reading the text, you are in direct communication with the author. This assumption leads to questions like, "In *Paradise Lost* Book 1, does Milton convince you that Satan is both attractive and corrupt?" This ghostly presence of the author is the final "authority" that can decide what the text means.

4. Simple evaluation

Once it is known what the author intended and so what the text means, it is possible to judge the text by how well the author achieved what she or he set out to do. This assumes that judging a work of literature is like judging someone's cooking: if you know someone intended to make chocolate chip cookies, you can judge them poor, OK or delicious. If you know what an author intended to do, you can ask questions like, "How successfully does Jane Austen show the growth of her female characters?"

While many forms of interpretation rely upon this idea of authorial intention, and it might appear to be common sense, it has been criticized for a range of reasons. These criticisms are now outlined.

Against authorial intention: The death of the author

Throughout this book I have argued that texts are always *interpreted* and open to different interpretations, stemming from readers' different worldviews. The idea that by uncovering the authorial intention it is possible to find out the "true meaning" or the "right answer" runs directly against this and underlies all the major objections to authorial intention.

1. Meaning: Is literature a code?

Is literature simply a code? Certainly, this is the impression given to many students: that a work of literature is about something – the theme – and that the job of the student is to discover what this theme might be. So is this really the case?

I would argue absolutely not for (at least) two reasons. First, the idea is self-contradictory. If literary texts were simply codes, then, paradoxically, literature wouldn't need to exist. Wouldn't it be much simpler to convey a message in a straightforward way rather than turn it into a work of fiction? Why write a novel to say "war is evil" when you could just say it, or go to a demonstration, or form a political party, or lobby (or even become) your own representative in government? Of course, there are texts with polemical messages, but when you respond to the message – for example, "imperialism is wrong" – it's the message or the argument you are responding to, not the work of literature itself.

But there is a more important reason why literature is not simply a code to be worked out. A code works like this: two (or more) people share a cipher where, for example, the letter "A" is represented by the number "1" and so on. One encodes, using the cipher, and the other decodes, using the same cipher. Thinking back to Figure 3.3 (p. 29), this cipher represents the same way of looking at a text, so both parties are agreed that 16, 15, 16, 16, 25, 1, 14, 14 is a name in code and not just collections of numbers. But, as I have argued, part of the point of literature is that it encourages different ways of looking at texts, creating different results. So, in fact, reading cannot mean *decoding* the secret message because there is no shared cipher, no one set of presuppositions we all share. Could you really see a text in the same way as a nineteenth-century author? Or even how your classmates view it? In having many different ways of looking, we have many different ciphers leading to many different "meanings."

2. Biographical evidence

This is also very much open to question. First, reading a letter or diary is not the same as interpreting a poem or novel. It would be interesting to find out what a text meant to its author, but that is not the same thing as thinking about what it means to you. Wimsatt and Beardsley, in their essay "The Intentional Fallacy" (1946), put it like this:

> In the spirit of a man who would settle a bet, the critic writes to [the poet] Eliot and asks what he meant [in his poem "Prufrock"] . . . [O]ur point is that such an answer to such an inquiry would have nothing to do with the poem "Prufrock"; it would not be a critical inquiry. Critical inquiries, unlike bets, are not settled in this way. Critical inquiries are not settled by consulting the Oracle.

Reading a text, interpreting a text, is not an activity that has a right or wrong answer. It is not like making a bet or solving a math problem. Unlike textbooks in the maths and the sciences, there are no answers in the back of your instructor's copies of the novels and plays you might read in class.

Second, whatever the "oracle" author said is itself another text open to interpretation. A letter saying "I intended such and such" is not firm evidence. Not only could it be a lie, plain and simple, but it is also open to interpretation because it is written within a certain historical period, where certain ideas were dominant, and because we, perhaps centuries later, may know things that the author didn't (and clearly vice versa). Authors might have very astute things to say about their own work, but what they say is only as valid as what another reader might say in determining the meaning of a text. Interpreting their work, authors are doing the same job as anybody else looking at a text. Another way of thinking about this is to ask, "Who owns words?" Wimsatt and Beardsley, discussing poetry, say that a text "is detached from the author at birth and goes about the world beyond his power to intend about it or control it." They argue that authors might shape language but that ultimately it is public property and readers may make of it what they will. This is not a modern idea: at the end of his long poem, *Troilus and Criseyde*, Geoffrey Chaucer (c. 1343/4–1400) wrote "go little book, go." He knew that, once created, the poem was out of his hands, and people were free to interpret it in any way they wished.

If an author's comments about intention are not authoritative, biographies are even less useful, being, after all, only an interpretation of somebody's life. It will certainly inform the reader about the author and her or his period, but it will not provide a "correct interpretation" for a literary text.

3. Authorial presence

Authorial presence is perhaps the most difficult assumption to understand. The question, "In *Paradise Lost* Book 1, does Milton convince you that Satan is both attractive and corrupt?" and others like it are, in a way, very confused. For they conjure up the rather worrying image of Milton appearing to you and arguing passionately that *Paradise Lost* Book 1 shows Satan as both attractive and corrupt. But surely, it is the *text* of *Paradise Lost* Book 1 and how you read it that would convince you (or not), rather than Milton himself? A text does not magically bring the author into the room with you – writing is just marks on paper. More than that, the very presence of the writing shows up the *absence* of the author. If the author were actually there, she or he could simply talk to you: the written text itself implies their absence, like an empty chair at a celebratory meal. (Look in this book, and others, at all the moments where the text says "As I have discussed . . ." or "We said earlier . . . . " In fact, none of these things are actually discussed or said at all; they are *written down*. Using the sorts of words that imply real speech is a way of suggesting that the author is actually there, present and talking to you. But this is metaphorical, not real. While you read this, I'm off somewhere else!)

Some critics argue that the author speaks *through* the text, but how could you tell when this was happening? In many novels or plays, several points of view are presented, for example through different characters. Which point of view is the author's? And even if there are passages written in the first person "I," how do we know if this is the author? It is with such questions that Barthes's essay "The Death of the Author" begins. He finds part of a novel where it just isn't clear who is speaking. Is it the author's voice? The voice of a role the author is playing (as the narrator, or as "the spirit of the age")? Is it always clear who or what is speaking? Is the author wearing a mask? Or, suddenly, does the "real" author appear? His point is that if you are looking for the "authentic" authorial meaning through a moment where the author "speaks," it is, in fact, very hard indeed to pin down for certain *where* on the page that moment is.

If writers are absent, how could we ever come to grips with the authorial intention? We can't ask them, and we can't even find out if there is a part of the text that was written to tell us "what they really meant." With the person irrecoverable, it seems foolish to try to work out his or her intention. Instead, perhaps, we should make what we can of the text.

4. Simple evaluation

Apart from the question of what you are to evaluate, if you cannot trace authorial intention, *how* should you evaluate? Who sets the standards? Does the question, "How successfully does Jane Austen show the growth of her female characters?" mean there is some fixed model of how successfully the growth of female characters *should* be shown? Or could you compare Jane Austen to another novelist of the period, Frances Burney (1752–1840), and judge who was better? The idea of judgment implies an objective neutrality that nobody could have and demands that everybody thinks in the same way. Even though it used to be thought that the job of the critic was to judge what great works were and who the great writers were, it is clear that judging a writer's "success" is more a result of the way the discipline has developed than a useful task in itself.

With these new ideas in mind, we could redraw the traditional diagram of the relationship between text and meaning as shown in Figure 8.2.

Authors, in saying what they meant by their work, can be seen as other readers, with an interpretation only as valid as that of any other person looking at the text. The author is no longer the all-important figure: The Author, as the saying goes, is Dead.

So why has the author always seemed so important?

Those who claim that the author is dead also look at how the figure of the author was "born," claiming this as another argument against authorial intention. The author and the importance of the role of the author in American

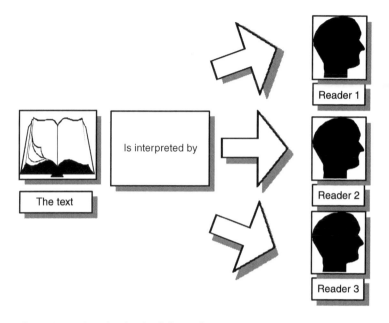

Figure 8.2 After the death of the author

and European culture was, like all ideas, invented. Of course, with broad concepts and categories of this sort, it is impossible to say exactly when it was invented, but it has been argued very convincingly that this idea of the author came into being in or around the eighteenth century. This is obviously not to say that people didn't write before this time but rather that their sense of identity as an author and their relation to their texts were different. Before mass printing, many medieval stories and romances, like *Sir Gawain and the Green Knight*, were without named authors (Chaucer is an exception) and who the authors were did not seem to matter so much. (In contrast, if present-day writers stay anonymous, it is precisely because it *does* matter who they are: they might want to escape persecution, or paying taxes, or scandal, for example.)

The concept of the author as the "true source" of meaning perhaps developed most fully during the eighteenth century: the period of the Industrial Revolution. During this time of massive change, writing became *property*, something that could be sold. It was possible to have a career as an author without having a patron, living by selling what one wrote. Since "ownership" of the words was important to generate income, the importance of attribution grew. If you find yourself at an airport bookstore, picking up something to read to pass the time on a long flight, you will probably look for a new book by a favorite mystery or science fiction or romance author. The name of this or that author (Stephen King, Walter Mosley Danielle Steel), functions like a brand, just as it would on a can of beans at the supermarket. [English major graduate and thriller writer Tom Clancy did in fact use his name to brand books as "Tom Clancy's Op-Center."]

Another major influence that helped to foster the idea of the author as the true source of meaning was the Romantic Movement – a loose collection of poets, thinkers, philosophers and writers in America and Europe in the late eighteenth and early nineteenth centuries. They focused on the created idea of the writer as *genius*, which didn't just mean "very intelligent" as it does today. A genius was a person whose immense creative and artistic power was a conduit between unseen powers (of Nature, for example, or the Imagination) and the world of human beings. Not only did this focus attention on the "author," the genius, but it became important to know who had this special ability and who didn't.

The Romantic concept of the author also stressed that an author must be original. However, some people have cast doubt on the very possibility of originality. Whatever original ideas authors might be trying to convey, they have only a limited number of preexisting counters – words – to use to do this, just as an artist has only a certain range of colors to paint with. Even new colors are only mixtures of old ones, and although the range of colors is wide – the visible spectrum – it is also limited (try imaging a totally different color that *no one has ever seen before*). Like colors, none of the words authors might choose are new: words are the only system of meaning that they can use. If authors want to explain what original idea they "mean," they can use only words that have preexisting meanings; so the words will already have *shaped* what the author can say. (This view reverses the normal assumption that an author shapes language: it suggests that, in fact, language shapes authors.) On top of this, much literature is bound by generic conventions, so any work has, to some extent, to fit an already established pattern. In a thriller, for example, the murderer can either be captured or escape. In a way, this doesn't leave much room for originality. These rules can be challenged and changed, of course, but this too relies on the rules inasmuch as rebellion has to rebel *against* something. These conventions are not part of the original intention of the author: the "original" ideas are reshaped by traditions of writing.

So the "author" is yet another invented category, and even the way this category is defined, as a person who communicates original ideas, is open to question. But what are the effects of this?

Consequences of the death of the author

If the author is dead and reading to discover her or his secret hidden intention is no longer the only logical course to take, there are new questions to ask. Perhaps one of the most important is how to understand the significance of the author today. The author might no longer be the source of meaning in a text, but it doesn't mean that the term has become irrelevant. Knowing about an author does still tell us some things about a text: the French philosopher and historian Michel Foucault (1926–1984) coined the term "author-function" to describe the way the idea of the author is used. For example, an author's name serves as a classification because you can be fairly sure what sort of text, broadly understood in terms of style and period, you will find under the name Louisa May Alcott or Stephen King. This is not to preempt the idea of meaning but to suggest that the name is used to group certain texts together. The author-function is also used, correctly or incorrectly, to ascribe value to texts. When, every now and again, somebody claims to have discovered a new Shakespeare poem, there is more fuss than when a new poem by a less famous poet is discovered. Again, if you like the work of a certain novelist, you might buy another novel by the same writer. The author's name also becomes a reference tag for other, often quite vague things like style or themes: critics discuss Margaret Atwood's style or Herman Melville's philosophy. Sometimes the names of authors are used as the tags for a whole series of "big ideas" – Darwinism or Marxism, for example. These ideas may have little (or even nothing) to do with those individuals in history, but the ideas still come under the classification of their name, so powerful is the author-function. In none of these cases is the author necessarily a source of authority on the meaning of the text.

Creative writing as a subject might also seem to "bring back" the author, and in some sense, it does because authors can be asked to explore their own work and explain their choices. However, as people work through the intricacies of writing, making choices about words or forms (a sonnet or a free verse poem? a novel or a short story?) or points of view (an all-knowing narrator? or a first-person narrator who is part of the story?), the idea that a single author controls the text that she or he writes seems more complicated as the language and genre itself start to shape the writing. The choices made in the writing – choices about the form of the text – seem to dictate how the writing proceeds and makes the author less and less powerful. And, of course, authors, when faced with their own texts, are readers too, making interpretations of the things they have written.

This has a corollary, too, in the publicity given to contemporary authors: interviews, lecture tours, public readings. Of course, we are interested in them as we are interested in all celebrities, but it seems odd because what we should be interested in is their work. It can sometimes seem as if we turn to authors precisely to give us the secret key to their work – an authoritative interpretation – that would save us actually reading the poems or novel.

Most importantly, the death of the author – or at least of the author's authority – leads to what Roland Barthes called "the birth of the reader." I understand this to mean that a literary work does have a meaning, but it isn't a puzzle or a secret to be found out, placed there in code by a genius author. Instead, it's something that grows as an interaction between the readers and the text itself. Each reader is able – or should be able – to interpret and to produce an array of different and stimulating meanings. You shouldn't be restricted by wondering what the author really meant. The meaning of a text lies not in its origin but in its destination: in you, the readers. Understanding a text isn't a matter of divining the secret but of actively creating a meaning.

Nevertheless, the author's intention is still endlessly referred to, sometimes to discount perfectly convincing and interesting readings of texts. It seems that many people want to find an authority to explain the text and provide the final answer. It is this wish for a final meaning that links the word "author" with the word "authority." This desire is particularly heightened in reading literature precisely because, I would argue, literature stimulates an unlimited proliferation of meanings. This idea, taken seriously, can seem quite threatening. If thinking about literature makes us think about the world, and there are no right answers about literature, are there any firm answers anywhere?

Summary

- It is often assumed that the author determines the meaning of a text. However, the reader also has a role to play.
- The conventional way of understanding a text as what the author intended makes a number of questionable assumptions about meaning, biographical certainty, authorial presence and evaluation.

- These ideas are open to question: we all read differently, and even authors can offer only an interpretation of their own texts. There is no one fixed meaning to be found or judged.
- The role of the author is an invention, developed in the eighteenth century.
- The term "author" still functions as an indication of style, genre or, perhaps wrongly, of quality. However, the meaning in the text relies more on your interaction with it than on the writer's intention.

9

Metaphors and figures of speech

- What is a figure of speech?
- What are metaphors, and how do they work?
- How do they affect us?

When you study a literary text, you often concentrate on the way it uses language and figures of speech. It is sometimes assumed that these figures of speech, and metaphors particularly, are just ornaments, there to decorate the texts and somehow show an author's skill. But they are much more important than this: they convey meanings at all sorts of levels, from the most mundane to the very deepest views we hold about ourselves and the world. Studying English involves not just appreciating figures of speech as ornaments but looking at and questioning their significance.

Figures of speech everywhere

As a rule of thumb, a *figure of speech* is the use of words or a phrase in a way that isn't strictly true; the words have been "turned away" from their literal sense and don't mean what a dictionary might say they mean. The technical term for figurative uses of language clearly reflects this: figures of speech of all kinds are called *tropes*, a term originating from the ancient Greek word *tropos*, meaning "turn, direction, course or way." There are lots of tropes/figures of speech, and they are not restricted to written texts: people use them all the time in everyday conversation and online. To show how widespread they are, here are seven everyday examples.

When you say something like "there were millions of people in the room," you are engaging in what is technically called *hyperbole*: an exaggerated statement that everybody knows is exaggerated. If you say "the book is really good" but mean you thought it was garbage, you are using *irony*, expressing your meaning by saying the opposite of what you actually mean (of course, people can misunderstand your irony). Synecdoche (pronounced "sin-ek-duh-key") and the closely related trope metonymy are two of the most commonly used figures of speech. *Synecdoche* occurs when people use a part of a thing to represent its whole. For example, the news reporter who says, "The White House has plans for the economy" is summing up the government in the image of the building. The sailor who shouts "Sail!" uses the word "sail" to stand for a whole ship. (Writers' names are one of the most frequent uses of synecdoche, and we hardly notice it. We say "Shakespeare" but mean "Shakespeare's works.") *Metonymy* occurs when the name of one thing is given to another thing with which it is associated. For example, "the pen is mightier than the sword" means that writing – an activity associated with the pen – is more powerful than fighting – an activity associated with the sword. *Animism* occurs when we describe something inanimate as if it had life; for example, "the angry clouds." (This is also known as the *pathetic fallacy*, in the old sense of "pathetic," which means roughly what we intend by "sympathetic" – sharing a feeling. A wild magic power, such as Elsa's from *Frozen*, which freezes things with ice and snow when you are angry or sad is a version of the pathetic fallacy: when the howling wind outside matches a whirling storm inside.) *Anthropomorphism* is rather like animism, but it names the trope that treats nonhuman things and animals as if they were human. The statement "my computer hates me" uses anthropomorphism (if we assume that only humans can hate). The British writer George Orwell's (1903–1950) famous satirical novel *Animal Farm* (1945) is anthropomorphic, as are the *Toy Story*, *Cars*, *Finding Nemo* and *Finding Dory* films – the characters (toys, motor vehicles, fish) behave as if they were human people. *Prosopopeia* literally means "giving a face to," and it refers to personifying things that properly are abstract. If you imagine death as a figure in black with a scythe, or war as a warrior, or justice as a blindfolded woman, then you are engaging in prosopopoeia. Because we use all these forms so often, you might begin to wonder if any phrases aren't figures of speech!

Metaphors in literature

The most widespread figures of speech, metaphors and similes, are of particular importance. Roughly, a *metaphor* is being used when we say that something *is* something else ("love is fire"), and a *simile* occurs when we say something is *like* something else ("my love is like a rose"). But how do these actually work? How do they convey meaning?

Like "trope," "metaphor" comes from the ancient Greek. It means "transfer," which is roughly what metaphors do. Formally defined in the *Oxford English Dictionary* as the "application of name or descriptive term to an object to which it is not literally applicable," metaphors transfer meaning by using a term to describe something else. George Lakoff and Mark Turner discuss this in detail in their book *More Than Cool Reason: A Field Guide to Poetic Metaphor* (from which many of the ideas in this chapter come). They argue that metaphors *transfer* meaning from one conceptual structure to another and so "allow us to understand one domain of experience in terms of another." For example, the first recorded use of the famous metaphor "the ship of state" was by the ancient Athenian ruler Pericles (c. 495–429 BCE). In it, the "state" from one domain (that of politics) is put together with the "ship" from another domain (the sea), so meaning is transferred from one to the other. The first time you hear the metaphor, you might wonder how the state is a ship and what they could possibly have in common, so you think about your concept of a ship. You might decide that a ship needs, say, careful handling during a storm, just as a state needs to be managed during a crisis. Metaphors make us think. Another classic example is, "Achilles is a lion." Achilles, the celebrated Greek warrior, is understood to be like a lion: very fierce and brave. Achilles is not actually a lion, but we understand his bravery using terms drawn from the natural world.

Traditionally, a simile is different from a metaphor because, where a metaphor says, "Achilles *is* a lion," a simile adds "like" or "as": "Achilles is *like* a lion." However, if we understand the process of metaphor as the transfer of meaning from one conceptual structure to another, there is actually very little difference between these two: they both work in the same way, and, because of this, the discussion of metaphor applies equally to similes. You could perhaps say that a simile is simply a weaker form of metaphor. It is less powerful to say something shares qualities with something else – "Achilles is *as brave as* a lion" – than it is to say it *is* something else – "Achilles *is* a lion."

In texts that we call literature, metaphors are said to defamiliarize language. The transfers of meaning they make are surprising or disturbing because the language with which we are familiar suddenly seems unfamiliar. It's this that makes us wonder what the text might mean. The novel *The Go-Between*, by L. P. Hartley (1895–1972), begins with the metaphor, "The past is a foreign country: they do things differently there." This metaphor works by using one conceptual domain, geographical space ("a foreign country") to describe another (time). But what does it actually mean to say, "The past is a foreign country"? That we can't speak the language? That we are lost? That we might not be welcome there? That we don't belong there and are only tourists? Is remembering events in one's life like being a tourist? It is this quality of *defamiliarization* that makes us think. Sometimes, of course, texts use metaphors that are so overused that they are clichés, which don't make us think at all. When we read, "My love is like a red red rose," we hardly notice it is metaphorical (it is technically a simile) because, since the Scottish poet Robert Burns (1759–1796) coined it, it has been used countless times. Roses have become a widely accepted metaphor for love, in literary texts and beyond. When Pericles first compared the state to a ship, his audience burst into spontaneous applause, but today the metaphor goes more or less unnoticed.

Metaphors in everyday speech

The examples so far have been broadly literary, but, as the rose-for-love metaphor suggests, we use metaphors all the time in our everyday speech. When we say that the "computers are down," we don't mean that they are literally down but that they are not working. Likewise, when the singer James Brown (1933–2006) tells us to "get down," he doesn't mean that we should lie on the floor but that we should start to dance. (On the other hand, when he tells us to "get up," he does mean, more literally, that we should get up and dance.) "Laid-back" people are not always reclining, and, even in the 1960s, "cool" people were not actually cold. In movies, gangsters "snuff out" people (you snuff out a candle, so the image is taken from a conceptual domain of lights and lighting) or "take them out" (leaving life is like leaving a building or being "taken out" of an equation). Even when we say that something important is "central," we are using metaphorical language, transferring meaning from a description of physical location ("central") to describe something's importance. In contrast to the defamiliarization that literary metaphors give us, these sorts of metaphors are often described as *dead metaphors*. We take their meaning so much for granted that we no longer even notice that they are not literally true. Indeed, "dead metaphor" itself is a dead metaphor.

Basic conceptual metaphors

Both our everyday metaphors and more explicit literary ones share a characteristic. They tend to build into, or rely on, what Lakoff and Turner dub the "basic conceptual metaphor" – the underlying metaphorical idea that generates a whole range of metaphors. This is best explained through an example, and Lakoff and Turner discuss at length the basic conceptual metaphor that says, "life is a journey." Many other metaphors, both literary and nonliterary, rely on the basic idea that "life is a journey." Lakoff and Turner show how "The Road Not Taken," perhaps the most famous poem by Robert Frost (1874–1963), relies on this metaphor: Frost sees his life as a journey down one of two roads. Many famous works of literature play with the comparison between life and a journey. For example, the great long poem *The Divine Comedy* by the Italian poet Dante (1265–1321) begins: "In the middle of life's road/I found myself in a dark wood." But this basic conceptual metaphor doesn't work just in literary texts. Pop songs use it all the time. Robert Johnson (1911–1938), the great blues guitarist of the 1930s, sang of "stones in my passway." In everyday conversation, we use the same basic conceptual metaphor all the time: we "go ahead" with plans; we get "sidetracked"; we reach "crossroads" and "turning points" in our lives; we do things in a "roundabout way"; like travelers, we are "burdened with things from our past"; there are obstacles "in our way"; babies "begin" the journey; and the dead "rest" at the end. This basic conceptual metaphor transfers meaning from one domain, our experience of journeys, to another, our experience of life. And we are so used to this basic conceptual metaphor that we all know what people mean when they say, "I have reached a crossroads" (even when, in fact, they are sitting quite comfortably at home) or "The path I must take is clear" (when no track is actually in sight).

Lakoff analyzes many other basic conceptual metaphors of this sort: "love is fire," for example, or "a human life is a year." (What season are you in? What season is the oldest person you know in?) "The past is a foreign country" would be an example of the basic conceptual metaphor "times are places," where different times are associated with different geographical locations. Each of these basic conceptual metaphors works like an engine for producing new metaphors that are generally understood. Interesting poems, novels and plays (and jokes, advertising slogans or, in fact, anything depending on metaphors) use these basic conceptual metaphors in new, defamiliarizing ways. They pull new things out of old models and shake up uses of language that we take for granted. Perhaps most radically of all, they can, occasionally, create new basic conceptual metaphors. These have the power to change the way we think about the world, and it is this that makes figures of speech and metaphors in particular so significant.

What metaphors mean and how they shape the world

So far I have suggested that metaphors are traditionally understood as the point where language "turns away" from its literal meaning. However, Lakoff and Turner's idea of basic conceptual metaphors changes this completely. These, like "life is a journey," are so deeply ingrained in us that, as Lakoff and Turner write in their book *More Than Cool Reason*, they are "an integral part of our everyday thought and language." Moreover, they have a unique, powerful and fundamental role in leading us to "understand ourselves and our world in ways no other modes of thought can." If life is a journey, then we can locate ourselves at a point on that journey – at the beginning, for example – and see things that happen in our life as things that happen to us on a journey – obstacles, crossroads, burdens and so on. The problem here is that this basic metaphor, which we so often accept without thinking, smuggles in a number of taken-for-granted ideas that, in general, we might disagree with if they were presented in another form. For example, if life is a journey, we might ask whether everything that stops you getting your own way is an obstacle. When we have a choice on a journey, it is often a choice of left or right, but in life, is it a choice of one of two options? Perhaps most powerfully, "life is a journey" smuggles in the idea that life must *go* somewhere, must have a final destination. Does it? The metaphor we choose to use to interpret the world in fact *shapes* how we interpret the world. How we look – the basic conceptual metaphor we choose on purpose or just fall into using by chance – shapes the events we are looking at. If you use a different basic conceptual metaphor, the events may look very different.

Another popular example of a basic conceptual metaphor is less personal and more political. Many people, including politicians and businesspeople, like to use the basic metaphor that "the country is a business." Once we are convinced of this, it allows all sorts of other decisions to be made that build on the same basic conceptual metaphor. An employee who is found to be stealing is usually fired. What policies does somebody who believes that "the country is a business" support when a citizen of USA Inc. is found stealing? Employees who are no longer able to work are made redundant. What would USA Inc. do for an "employee" who could no longer work? Again, the question is which basic conceptual metaphor you decide to use. If you chose "the country is a family" as a basic conceptual metaphor, you might come up with very different answers to these questions (who counts as "in the family," for example).

These basic metaphors control our view of the world. Part of majoring in English involves understanding their power and analyzing them when they occur, to find out exactly what they take for granted. We might begin with literary texts, certainly, but this can stimulate you to look for basic conceptual metaphors in the wider world and to think about the ideas they depend upon. This is part of thinking about the context in which texts are written and understood.

But even more than this, part of the point of studying English and perhaps part of the point of literature is to offer new metaphors – not just to surprise us by defamiliarizing our normal use of language but to offer whole new ways of conceptualizing the world. For example, it is common to think of knowledge as a tree: there is a trunk (core subjects, perhaps) and each subject is a branch, subdividing into smaller branches and twigs as the subject becomes more specialized and farther away from the trunk. But what if a better and more interesting metaphor for knowledge and learning was not a tree but, as the French philosopher Gilles Deleuze (1925–1995) suggests, a rhizome (a plant such as grass, a potato or bindweed)? These plants have no center, no core subjects, but move, grow and change independently of a central authority. Each subject, each clump of grass, would be interdependent, not a refined speciality relying on others. There would be no core subjects that everyone had to acknowledge but an array of different and equally valid sorts of knowledges. The point is that basic conceptual metaphors help determine the sort of things we think and that doing English helps us to explore and to question these metaphors.

The huge question still remains of *why* we think in conceptual metaphors and why figures of speech are not just ornaments. I suggested at the beginning of the chapter that a figure of speech was "the use of words or a phrase in a way that isn't strictly true." Once, in a lecture, I asked what the literal meaning of the metaphor "He's cool" is. Somebody shouted back, "He's hot!," which made everybody laugh, as one metaphor was simply replaced with another. But there was a point here. The idea of "coolness" can be understood only metaphorically – there is no literal truth behind it, no actuality that could be unambiguously pointed at, no simple truth from which language could "turn." The German philosopher Friedrich Nietzsche (1844–1900) went further. For him, there was no literal truth at all: truth is made up from metaphors. He wrote:

> What, therefore, is truth? A mobile army of metaphors, metonymies, anthropomorphisms: in short a sum of human relations which become poetically and theoretically intensified, metamorphosed, adorned, and after long usage seem to a notion fixed, canonic and binding; truths are

illusions of which one has forgotten that they are illusions; worn-out metaphors which have become powerless to affect the senses.

For Nietzsche, there is no literal truth that language can convey, and all words in language are really just figures of speech. We think in metaphors, and they grasp and control our minds. One contemporary philosopher, Jacques Derrida (1930–2004), describes the way metaphors grasp our mind as "metaferocity." We are so used to metaphors and what they take for granted, however, that they have become dead metaphors – worn out and powerless – and yet they seem to be true. If Nietzsche is correct, the usual understanding of tropes is completely incorrect: they are not "turnings" of language away from truth or just ornaments. Quite the opposite – figures of speech are the fabric from which the truth of the world is made up. So when we look at the metaphors in a poem, we are, at the same time, thinking about how the whole world is understood.

Summary

- Figures of speech (or tropes) occur when language is used in a way that isn't strictly true. We use them all the time. Metaphors and similes are the most common examples.
- Metaphorical language describes one thing as another ("my love is a burning fire"). It works by transferring meaning, allowing us to understand one domain of experience in terms of another. Literary metaphors defamiliarize language, but we also use metaphors in everyday speech, often without noticing. A metaphor we don't notice is a dead metaphor.
- Both literary and everyday metaphors rely on basic conceptual metaphors, such as "life is a journey." These work like engines for producing other, generally understood metaphors. Some people argue that these basic metaphors are so deeply ingrained in us that they shape how we see the world. The philosopher Nietzsche argued that metaphors are so taken for granted that we simply think of them as the truth and no longer recognize them as metaphors at all.

10

Narrative and closure

- Why are stories important?
- How do we understand narrative and narrators?
- What is closure?

The last chapter looked at some of the ways in which language was important in texts and in understanding the world. But we also use stories to order the world, and, of course, story – or, more technically, narrative – is what literary texts are made of. A lot of newer approaches to literature draw heavily upon questions of language and meaning, as discussed in the previous chapter. However, there is also a great deal of work done on the way we use narrative to order and give meaning to our world(s). Studying English means engaging with and understanding narrative.

How are narratives made?

Part of the reason many people study literature is that they are swept away by stories. They find books "unputdownable" and read late into the night. This is part of the power of narrative. It isn't just novels, of course. Soap operas, poems, plays, films and, in fact, nearly every sort of text rely on this narrative drive, the desire to find out "what happens next." Narrative is everywhere. It isn't only in fiction: it is also central to each of our lives. When we are born (or at least when we can take notice of what's going on!) we find ourselves thrown into the middle of things and other people's lives. In order to make sense of what's going on, we tell ourselves stories. I'm sure that most of us have – when we were younger – done the same thing as the main character in the Irish novelist James Joyce's (1882–1941) book *A Portrait of the Artist as a Young Man*. At school, he wrote down "his name and where he was: Stephen Dedalus, Class of Elements, Clongowes Wood College, Sallins, County Kildare, Ireland, Europe, The World, The Universe." This is a sort of story, about identifying who and where he was. And we are told and we tell ourselves stories about what has happened to us or who we are and who we want to be, whether we write them down in diaries or transform them into poems or fiction or – more often – just think and talk them through. These stories are how we organize and sort out the chaotic world. They are how we give the world meaning. For example, if you are asked to tell someone about yourself – at an interview for a job, say – you quickly outline the broad story of who you are, where you come from and so on. You probably would not relate some anecdote from your childhood or describe in agonizing detail your morning's journey. You present your story of yourself, organizing the narrative and selecting what you take to be the most effective and meaningful facts. This "organizing" is a way of *constructing* the story of yourself: an activity that everybody undertakes, consciously or not.

Because narrative is so important and so all-pervasive in our lives and in the texts we read, study, watch and create, critics have been trying to define and understand it for a long time. A Russian theorist, Vladimir Propp (1895–1970) studied folktales and argued that each of these fairy stories was put together using some or all of thirty-one narrative motifs. In each story, for example, the hero would leave home or the story would end in a communal coming together – a wedding or reunion. In a way, Propp was making an attempt to make more formal what, in the case of fairy stories, is quite obvious: that Prince Charming from "Snow White" serves the same function, or does the same job, as Prince Charming from "Cinderella." His "character," as such, doesn't really matter – as long as he is charming, of course. (Bill Willingham's comic *Fables*, in which fairy tale characters are real, makes great play of this: the same devilishly handsome but unreliable man is the ex-husband of Cinderella, Snow White and Briar Rose.) What matters is his role in the story as the hero, the rescuer of the heroine from the villains, whether they are the evil witch or ugly sisters. The villains, too, are significant only because of their function. Hollywood blockbusters can be analyzed (and indeed are often written) in the same way: function by function. Propp's ambition, as well as the aim of those who came after him, was not simply to analyze fairy stories or movies but to create a "science" of narrative, or narratology. This would be a way of breaking down all narratives in all forms into their constituent parts. There have been many rigorous attempts to do this.

These attempts have had some useful results. Certainly, this "scientific" approach to narrative generated some quite precise terms. For example, one term that is quite widely used is *diegetic*: this is a more formal way of specifying the "world of the story." Gotham exists in the diegetic world of Batman; East Egg and West Egg exist in the diegetic world of *The Great Gatsby*; Bikini Bottom exists in the diegetic world of SpongeBob. However, overall, this project seems flawed in a number of ways. In looking for general laws of narrative, narratology passes over the individual nature of specific texts and is interested only in functions. Although this is fine for simpler forms of narrative (fairy tales, or rather formulaic fiction, say), it seems to miss out much that is valuable. It is also blind to historical difference: because both Shakespeare and SpongeBob rely on narrative, so narratology treats these two art forms from different periods, media and genres as if they were the same. Moreover, each attempt to analyze the text has presuppositions: Propp chose to look at "the actions of characters," while another narratologist, Gérard Genette (b. 1930), chose to look at the ways a novel uses time, exploring how the action jumped backward and forward (with flash backs, or what he called *analepsis*, and flashforwards, or *prolepsis*). However, these approaches are neither natural nor scientific but arbitrary. One could analyze a novel in terms of *where* the narrative takes place and how the characters move, rather than *when*, for example.

However, this approach did focus interest on some specific issues. Perhaps the most important of these is the question of narration; that is, who is telling the story?

Narrators

Although not all stories have narrators, most do, and most literary texts certainly do. Sometimes the narrators are inside the diegetic, that is, inside the world of the story. Often these are first-person narratives in which "I" tell the events that have happened to me, or voice-overs in a film. *Heart of Darkness* by Joseph Conrad (1857–1924) is a first-person narrative. But if one person is telling you a story, this immediately raises questions of narrative reliability.

How far can you trust the narrator of a first-person story? After all, the narrator doesn't know what the other characters are thinking or doing, and they are surely telling you what they think is important and what they think is going on. It is easy to imagine a novel in which what the first-person narrator tells us is going on and what is actually happening are different. Indeed, in *Heart of Darkness*, there are "slips" between the narrator and the story he is telling, leading readers to suspect that the first-person narrator isn't being entirely honest. For example, Marlow, the main narrator in *Heart of Darkness*, describes the genocidal activities of the Europeans in the Congo and yet seems unwilling to face – clearly – his own complicity with them. (Perhaps it is precisely this struggle between his need to confess and his unwillingness to do so that gives the novel its particular and peculiar tension.)

Sometimes, however, narrators are outside the diegetic and seem almost to disappear from the story they are telling. This is third-person narration, in which characters do things that the (nameless) narrator describes. These sort of narrators are often described as omniscient, all-knowing, because they do seem to know everything. They can, quite literally, get into the heads of the characters and tell you whether they are afraid or happy or if they are hiding something. Traditionally, most novels have this sort of narration: it is the narrator who begins *Pride and Prejudice* by Jane Austen (1775–1817) with, "It is a truth universally acknowledged that a single man in possession of a good fortune, must be in want of a wife." (But is the narrator being *ironic*? Even omniscient narrators are characters.)

However, omniscient narrators do not narrate everything. Indeed, part of the power of narrative – first- or third-person – lies in the *focalization* of the narrative on or with certain characters. For example, a detective in a murder story may not be the narrator, but the narrative accompanies her or him, rather than, say, one of the bystanders, and reveals only what the detective finds out. The narrative often takes one person's side over another, for example, and who or what the narrator focuses on is very revealing. The story of a murder and its solution told from the point of view of the murderer would certainly be an interesting read, but would it be a detective story? The use of focalization is very significant, too, in the way it can change what we think about a story. For example, in an adventure story, the narrative focuses on the hero, not on, say, the hapless guard shot dead by the hero in the third chapter. But if the focalization showed that guard to be a character too, with his own hopes and fears, perhaps the story – and the murderous hero – would look very different. Perhaps the guard works for the villain because he needs the money for an operation for his ill mother or because he comes from a poverty-stricken part of the world, and the hero simply guns him down. These choices of narrative focalization shape the meaning of the text. Thinking about focalization in this way can offer unsettling insights into texts. For example, although *Heart of Darkness* is set in Africa and is in no small part about the colonial relationship between Africans and Europeans, and even though there are African characters, at no point is the narrative focalized through an African character: the reader is never allowed "inside their heads." Once this is clear, then other aspects of the novel and its context come more clearly into view.

Closure

However, these ways of analyzing and understanding narrative don't seem to explain either its importance or why we find narratives so compelling. The simultaneous explosion of both nonsynchronous TV watching and social media offers an insight on narrative and how it works. You have seen a warning for a spoiler alert on a social media post or episode recap. "Spoiler alert" is shorthand for, "Don't read this if you don't want to find out what happens on this episode before you watch it." However, this idea of spoiling raises the question of how we value stories. Would we still read *Moby-Dick* if we knew beforehand (spoiler alert!) whether Ahab or the whale would come out on top? Thinking about what is and is not spoiled in your experience by a spoiler can be a useful way of thinking about narrative and reveals something crucial. Wanting both to know how things end and to experience the process of getting to that end, the putting off of the finish, is absolutely vital for all stories, from the longest Victorian novels to the shortest cartoons to the most recent HBO series. It is one of the most important things about literature (and about life), and it is called *closure*.

Here's a test to see how significant closure is: would you read a book or see a film if you knew that the end was missing? Maybe not. This, as well as the phenomenon of spoiler warnings, shows not only how important closure is but also – oddly – how the ending of a story is already implicit in its beginning, because even right at the beginning a story has an end, a goal, a conclusion in mind. Stories are *teleological*: they begin with an end in mind. Despite the fact that it is so important, closure – this wanting to know how things end, this feeling of teleology – is quite hard to pin down and, like many things in English, it is often just taken for granted. (For example, the basic conceptual metaphor discussed in the last chapter – "life is a journey" – is a teleological metaphor: suggesting that a life is a journey implies that there is a final destination that will provide closure for "life's story.") However, although we may not be able to offer a strict definition of our desire for a "sense of the ending" (as a famous book by the British critic Frank Kermode named it), it is possible to see closure at work. Characters on TV often say that they need closure in their life ("I need closure in my relationship with my ex"). They mean that they need to finish one part of their life, tying up all the loose ends so that they can move on. And in everyday life, many events are "acts of closure." Graduation day at the school or university is an event that marks the end, the closure of a part of your life. Funerals don't end the mourning, but they end the immediate period of grief after someone has died. We feel that these acts "close a chapter." This is a key idea in understanding closure: we *construct* closure just as we construct narratives of ourselves.

This is why closure is also one of the links between the stories we read or see and our own lives: they both rely on closure to make sense. This is how the novelist Henry James (1843–1916) put it:

> Really, universally, relations stop nowhere and the exquisite problem of the artist is eternally to draw, by a geometry of his own, the circle in which they shall happily appear to do so . . . He is in the perpetual predicament that the continuity of things is the whole matter for him, of comedy and of tragedy; that this continuity is never broken, and that, to do anything at all, he has at once intensely to consult and intensely to ignore it.

What this means is that things ("relations") keep on going on (your school or college will still be there after you leave) and what artists do is decide how to "draw" the events, what bits to choose and where to end. In this sense, we are all artists whenever we tell a story. Life goes on with its continuous events, and all stories, fictional or real, are made by decisions about where to end. Do you stop telling the story at the couple's wedding (a happy ending)? Or at the new husband's fatal car accident (a very sad ending)? Or when the widow falls in love again (happy) but is brokenhearted when her new love leaves for Brazil (sad)? And so it goes on. For Henry James, the artist's choice is where and how to end, how to construct a sense of closure.

Nearly all texts use closure – it's how all stories work after all. Some end with a very satisfying tying up of loose ends: the murderer is caught, the wedding happens, or the main character dies. Others are less final, and, in recent years, some texts have purposely tried to avoid closure by leaving lots of loose ends, unfinished plots and so on. An example of this is the short novel *The Crying of Lot 49* by Thomas Pynchon (b. 1937). This novel in part concerns what might or might not be a massive conspiracy that the central character investigates, but she never actually discovers whether it is a conspiracy or whether she is imagining it all. The point is to show that closure is sometimes a bit hackneyed and to draw attention precisely to the way in which it is constructed, especially in fiction. But these sorts of "anti-closure" novels work only because you expect closure, so they, too, are using closure but in a different way.

"Closure" is a key term for studying literature and is vitally important for thinking about how the content, form and structure of a text shape its meaning. Knowing about closure allows you to compare different novels, poems and plays by looking at how they achieve closure, which means, really, how their stories work. You might ask yourself how a thriller achieves closure compared with a historical novel, for example, or how conclusive the ending actually is.

The question remains of why our need for closure is so strong. Why do we want to know what happens? There is no one answer to this, but much of the critical theory you will encounter in higher-level English studies will examine the issue. Frank Kermode suggests that, perhaps, because we know how the story of our own life is going to end – the way all lives end, in death – the endings of the stories we read or tell are like "little deaths" that will allow us to come to terms with our own, real death. It might also be because closure is a way of constructing and imposing order on a chaotic world, a way of drawing our own circle.

Of course, English is not immune to the desire for closure. The desire to come up with the right answer or the final interpretation is in part the desire for closure: the end of studying this book or that play. However, if closure is constructed, so are our readings and interpretations that rely on closure. Closure is unavoidable and necessary but, at the same time, has to be seen for what it is: a way of finishing a story or – in the case of the subject of English – finishing with a story. But literature, although perhaps based on closure, has a habit of not letting itself be closed up. Because there are many different interpretations of texts, we know that a sense of closure – of a final answer – is a construct and that an interpretation can be opened again, reconstructed or deconstructed by different readers in different contexts. So to think about closure is to think about our own reading and the ways in which we impose our own meaning on texts. When you're majoring in English, the texts might be literary, but we can also think about the ways in which we impose meaning on the "texts" of our lives and those of others.

Summary

- Narratives are everywhere and are very powerful. We tell stories in order to make sense of the world and ourselves.
- Some critics have tried to develop a science of narrative, which has helped refine discussions of how stories work.
- There are different sorts of narrators. Some narrators are inside the story, and others are outside, but both sorts of narration shape the meaning of the text.
- Closure is our sense of an ending and part of all narrative. Some critics suggest that we seek closure to impose order on our lives or to come to terms with our own deaths.
- In thinking about closure, we can also think about the ways in which we try to impose meaning and final interpretations on texts.

11

Creative writing and critical rewriting

- What is creative writing?
- What is critical rewriting?
- How is creative writing marked?
- How has creative writing changed doing English?

When I was a student, I asked a question about sonnets: the lecturer – a poet as well as an academic – astonished me by saying that the best way to really find out about sonnets was to write one. I think I was astonished both by the audacity (that I might dare to write a sonnet, as if I were an author and not only a literary critic, a student!) but more by the obvious simple rightness of this suggestion. After all, no one learns to swim on dry land. In English, though all reading is active, creative writing particularly stresses heuristic learning: *heuristic* means learning by doing, finding out, rather than being told or by simply analyzing.

What is creative writing?

The disciplines of English and of creative writing are clearly related: both involve writing and responding to literature and, as Gerald Graff's history of English, *Professing Literature*, showed, many famous critics who helped shape English as a subject were also poets and novelists. Like many siblings, they do not always get on so well, but they can learn from one another.

In a general sense, all the texts studied in the discipline of English are "creative." But creative writing more specifically names the study not only of the works of past and current writers by students but also of the production of new creative works. Students become *writers* not only of interesting essays and analyses of texts but of their own poems, short stories, novels, plays, journals and other literary forms. But what is creative writing? Can people be taught to be poets, playwrights or novelists? Like most questions in English, answering this involves several other ideas. If creative writing is about enabling students to be authors, this too raises a question, one that has attracted a great deal of contentious discussion in creative writing about what an author is or does.

Some of the appeal of creative writing lies, of course, in the figure of the "author" (which is one reason why the issues discussed in chapter 8 about authorship, authority and intention have been so contentious). There is an attraction in the idea of being an author, becoming famous or, perhaps more interestingly, being able to articulate something important about oneself or the world, to *say something*. But this idea of the author, as a sort of inspired genius, absorbed in self-expression, has also been a problem for these courses. As I suggested in chapter 8, the idea of what the author was has changed over time. To some extent, the idea of authors as inspired figures, able to express the truth about themselves or the world, is inherited from the Romantic period at the beginning of the nineteenth century. This idea of the author fits very well with current unquestioned ideas about celebrity, in which it seems that famous writers, actors and musicians and so on are assumed to know not just more about writing, acting and music than noncelebrities (which is reasonable) but also more about world politics, life and everything else. But if one believed completely in this idea of what authors were or are – inspired geniuses – then, of course, creative writing could not be taught and would not be part of English. It would just "happen" to you. In fact, of course, all writers have learned to write. Writing is not a natural skill: like learning an instrument, it has to be taught, practiced, developed and so on.

In response to this, then, a contrasting view of creative writing is that it is a discipline that exists as a "craft," to teach the student the technical skills of writing. In this context, Steve May, in his book *Doing Creative Writing*, answers the question of whether creative writing can be taught very clearly. He begins:

You want to write a story. I don't know what your story is about, what genre you want to write it in, or anything else about it. However, because (I believe) stories share common features, I can be fairly certain that if you ask yourself the following questions about your story, it will help you discover its shape and point:

1. Who is it about?
2. How do they change over the course of the story?
3. What do they want?
4. Do they get what they want?
5. Who (or what) is trying to stop them?
6. What are the key events in the story?
7. Which event decides whether the main character gets or doesn't get what he or she wants?
8. How does it end?
9. Which other characters are absolutely essential to the story?
10. Where is the best place to start?

These are technical questions about writing and the decisions writers make. In this view, the author works to learn a craft, to put together a story, poem or play and to learn the elements that make them up. While this is empowering, however, it also implies a vocational sense: that someone doing creative writing will end up making a living as a writer. However, not all creative writing students will make a living as writers, and, of course, many students taking a creative writing module in higher education may not want to become writers, so it's not correct to see these courses as just professional training.

In her book on the subject, Michelene Wandor writes that creative writing is a "mode of imaginative thought." Here she echoes the wonderful British novelist Angela Carter (1940–1992), who writes what "I really like doing is writing fiction and trying to work things out that way." For Carter, novels were partly "thought experiments" that tried to explore imaginatively – but not necessarily answer – certain problems or issues, "to work things out." However, all subjects one learns and studies are attempts to explore certain problems: physical geography explores why landscapes are the way they are, for example, and the sciences, at their best, need creative and imaginative responses. What is at issue with creative writing is not just that it is a mode of thought but *how* it thinks, as it were – what its tools are.

The tools of creative writing as a subject are the literature, criticism and theory that are involved with any literary creative act. (Angela Carter again: "reading is just as creative an activity as writing and most intellectual development depends on new readings of old texts. I am all for putting new wine in old bottles; especially if the pressure of the new wine makes the old bottles explode.") Creative writing is another important way of engaging with literature, another of the new ideas that are reshaping English as a subject, stressing the *heuristic*, learning by doing. It brings together you, your ideas, what you have read, different and new thoughts and interpretations with hands-on experience of writing in a range of different ways.

One of the techniques that English has learned from creative writing and perhaps the most widely practiced in English courses at high school and university level is called *critical rewriting*. This involves taking an established text and reworking parts of it: for example, a scene from a novel could be rewritten from a different perspective or in a different context. This is not different, really, from what theater directors or film and TV adaptors do when they stage a Shakespeare play ("What happens to *Hamlet* if we are performing it for GIs?"; "What happens to *The Taming of the Shrew* when we move it to Padua High School?"). More interestingly, this process of creative rewriting can tell us about the original text, about the new format and about their interaction. For example, the Facebook status update version of *Pride and Prejudice* tells us about the Jane Austen novel by focusing on the key incidents, on contrasting views and on the events of the narrative. But it also tells us about the genre of Facebook Status Updates: it shows, really clearly, for example, the discipline involved in having to say something funny, clever or moving in only a few words. Most interestingly, both the original novel and the Facebook rewriting – and Facebook and Twitter more generally – share a common feature in their use of irony and humor. Mallory Ortberg's (b. 1986) *Texts from Jane Eyre* (2014) does a similar thing for a wider range of literature. If people learn new things by reading texts in different ways, they also learn by writing or rewriting them in different ways.

Creative writing as a subject, then, although it's been around for a while, is one of the new ideas that forms part of doing English today. If some of the ideas described as theory focus a little more on *what* texts mean, creative writing focuses on *how* they mean, even though, at some deeper level these two – what something means, how it means – are versions of the same question. In an excellent book about new ways of teaching English and creative writing, Ben Knights and Chris Thurgar-Dawson argue that the value in writing lies not so much in self-expression or Pulitzer Prizes but "as an educational process that permits deeper engagement with the already written."

They go on to argue that, although the new ideas in English utterly changed the subject – thus, the need for this book – the way the subject was taught and what students actually wrote have not yet been changed. Creative writing and critical rewriting are about changing not only what students think but also what they do, not only to produce essays but also different sorts of heuristic responses to literary texts, ideas and thoughts. In this, the creativity of writing and the creativity of reading are really revealed as the same thing: the best creative writers are also the best creative readers. Creative writing is active reading. In this way, creative writing and critical rewriting are changing the nuts and bolts of how English is studied and taught.

Nuts and bolts and assessment

Changing the nuts and bolts of English means, among other things, changing how work in English and in creative writing is graded, and one of the questions that has often been asked about creative writing, especially in education, is how it might be assessed.

On the one hand, "creative writing" is assessed and valued all the time: newspapers or Internet reviews judge novels, poems and plays; people who teach courses choose the most interesting books they can; we recommend (or don't) books or TV programs to our friends. As I suggested in writing about the canon in chapter 6, you can't escape literary value. On the other hand, it is very hard to imagine giving actual marks to great poets. People involved in creative writing and running MFA writing programs have spent a very long time thrashing out these issues, and the answer again turns on the heuristic and practical. To begin with, much of the work done by students in this area is divided into two parts: a creative part – the poem, short story, film scene – and a critical, or reflective, part.

Perhaps ironically, in order to assess the first, creative part, teachers have provided very detailed assessment criteria, focusing on relevant issues to the writing:

- How effectively did the piece presented explore the point of view of the main character?
- How clearly has the plot been advanced in this scene?
- Does the choice of rhyme scheme (or lack of rhyme scheme) suit the subject?

In turn, the critical and reflective part might be, for example, an essay on the technical features (say, the sort of tropes that, following chapter 9, your creative piece might have used and why you chose them), or a piece laying out the aims of your creative work and assessing how far you were successful, or a discussion of how you saw your creative piece about growing up in Tennessee in the wider context of other novels and stories about childhood.

If the main form of assessment for creative writing has been the creative work together with some reflective work, the main form of teaching creative writing has been the workshop or seminar. The novelist and creative writing teacher Douglas Cowie, from Chicago, writes that a

> Creative Writing Workshop is usually run in part or in whole as an opportunity for student writers to submit work-in-progress for consideration by their peers, under the guidance of an experienced writer. A student's work will be read by each member of the group, either during the workshop or beforehand, and the group will then discuss the poem, story, novel extract or play, offering constructive criticism about how the piece does and doesn't achieve its aims, and also offering suggestions for improvement of the piece. There are a number of variations on how creative writing workshop discussions can be run, and in my own teaching I've organized them in different ways depending on (among other things) the experience, size and ability of the particular group. However, the basic foundation of a student's work-in-progress receiving oral and/or written feedback from his or her peer group under the guidance of a more experienced mentor is common to most workshops.

Like much in English, the creative writing workshop is controversial and can, of course, be difficult. However, workshops do also reflect an interesting and often forgotten thing about writing itself. While the image of the writer is of a lone person creating by themselves – and there are writers that fit this stereotype – usually writers are involved in communities of different sorts. Novelists and nonfiction writers have friends and competitors (often the same people) to whom they show drafts, as well as editors (*wonderful* editors, as the Routledge editor Polly Dodson suggests I add here) and publishers who look at their work and make suggestions. Writers for the stage, TV and film have many collaborators from (one hopes) the informed and sympathetic directors to (one fears) the less scrupulous people for whom art is simply another business. Poets read to each other, attend events and festivals, form supportive groups, compose manifestos and make aesthetic friends and enemies to help shape their own writing. The creative writing workshop is a version – perhaps a delimited and controlled version – of these different and jostling communities, so doing creative writing is also about learning how to be in these communities.

Creative English?

Creative writing/critical rewriting should enable you as writers and allow new ways of understanding literature and its interpretation. But these changes have an impact on all the other parts of the subject of English too. Paul Dawson in his book *Creative Writing and the New Humanities*, for example, discusses "Fictocriticism" or "personal criticism," a sort of criticism that draws more and more on the personal voice and on personal responses to texts. But, at its root, all responses to literature are our own, reflected on, evaluated and changed by all we have learned and thought. You might not need to invoke a personal voice or (as the eminent Shakespeare critic Stephen Greenblatt does) tell a story about one's father to be creative in your responses. Another great twentieth-century critic, Geoffrey Hartman, argued as long ago as 1980 that "all criticism entails a rethinking, which is itself creative . . . in every aspect of learning and life." Because of this, like creative writing, other sorts of ways of responding to texts, other sorts of literary criticism "may cross the line and become as demanding as literature"; criticism is "an unpredictable and unstable genre" not limited to a commentary on literary texts. All responses to literature are creative.

Summary

- English and creative writing are related disciplines.
- Creative writing focuses on you as a writer.
- A creative writer is not an inspired genius or simply following a craft but someone learning about and responding to literature in a different, more heuristic way.
- One form of this is critical rewriting, which involves reworking already existing text or part of a text.
- Creative writing is often assessed by both a piece of creative work and a piece of reflective work.
- Creative writing stresses the creativity at work in all responses to literary art.

PART IV
ENGLISH AND YOU

12
English, identity and politics

- Why is English involved with national identity?
- What is cultural heritage, and what does it have to do with English?
- What does English have to do with politics?
- How do different critical attitudes approach the issue of literature and politics?
- Why has English been a political battleground?

In English as a discipline, ideas about individual, communal and even national identity and about society, culture and politics meet, mix and often become indistinguishable from each other. This chapter explores the complicated interaction between English, identity and politics and seeks to explain why English is often a battleground where different groups clash.

Where does your communal identity come from?

It was once commonly believed that your communal identity or nationality was in some way part of your body – in your genes, bones or "in the blood." But few people now believe this. If it were true, it would be impossible to change nationality. Since identity is not in your blood, it cannot be defined by race or ethnicity, as is sometimes suggested. (Indeed, many people now argue that the idea of defining and categorizing somebody principally and exclusively by race arose in the eighteenth century. This is not to say that differences between peoples weren't previously noticed but that these differences weren't seen as summing up all a person was. This idea of "race" was used to support the growing colonial expansion of European nations and also to "justify" the unjustifiable evil of slavery.) Benedict Anderson's (b. 1936) book *Imagined Communities* (1991) offers a crucial insight into communal identity and the idea of a nation. He argues that communities are created, or constructed, *culturally*. They are "imagined communities" because they are constructed first *in the imagination*. There is a shared stock of images, ideas, stories and traditions, all of which go together to help people "imagine" (and so identify) themselves as American or as members of Hispanic-American, African-American, Native American or other communities. These communities imagine themselves to be a *community of people who share something*. An imagined community binds people together in a "we." In the case of a national identity, nations foster the idea that all those who claim to be of that nationality lay claim to something in common. This idea of community implies a deep comradeship, crossing over boundaries of class, race, gender, education, upbringing, religion and so on.

As I discussed earlier, the origins of English as a subject of study was bound up with nationalism, in the UK and the USA. Almost as soon as the United States became a nation, its literature betrayed an interest in forging a distinctly American and not English identity. *The Contrast* (1787) by playwright and politician Royall Tyler (1757–1826) opens with a speech that begins:

> EXULT, each patriot heart! – this night is shewn
> A piece, which we may fairly call our own;
> Where the proud titles of "My Lord! Your Grace!"
> To humble Mr. and plain Sir give place.
> Our Author pictures not from foreign climes.
> The fashions or the follies of the times;
> But has confin'd the subject of his work
> To the gay scenes – the circles of New-York.

In spite of Tyler's effort, American concerns about the relative worth of American literature persisted, exacerbated in 1820 by Sydney Smith's famous comment, "In the four quarters of the globe, who reads an American book?" Some years later, the Irish novelist George Moore (1852–1933) observed:

"James went to France and read Tourguénieff," he wrote. "W. D. Howells stayed at home and read Henry James."

Making things more complicated, although the connection between the English language and the English people is relatively straightforward, the United States is a polyglot nation. Concerns about the links between American identity and the English language have a way of popping up in laws that propose designating English as the official language of a state or town – usually in response to a growing population of U.S. citizens who speak a language besides English as their primary language.

These ideas about identity affect how you behave, your expectations, your relations with others and, more importantly perhaps, others' relations with you. It affects how you understand the world and your place in it – it affects the presuppositions you have when you read. Your communal identity, in no small way, *makes up who you are*. And because this identity is imagined, it is actually made out of cultural ideas and images. It is usually assumed that your identity creates your culture, as if the tree of culture grew from the soil of identity, but, in fact, *it is your culture that creates your identity*. Culture is a vital component – if not *the* vital component – of identity.

In his book *Keywords* (1976), the influential Welsh critic Raymond Williams (1921–1988) wrote that culture is "one of the two or three most complicated words in the English language." The word has at least three different but interwoven meanings. The first is personal: to be cultured is to have undergone a process of learning and development (to be, as some of the founders of English might have phrased it, "civilised"). The second meaning refers to culture as high culture: the great (that is, canonical) works of literature, opera and classical music, for example. The final meaning refers to culture as a word to sum up a much wider array of things: images, objects, pictures, comics, "pulp" literature, religious ceremonies, pop songs, films, clothes, television, soaps, team histories, traditions and everything else that goes into making up the world we experience. This meaning of "culture" makes up what Homi Bhabha (b. 1949; a leading contemporary thinker on culture and identity) calls "the scraps, patches, and rags of daily life," all the made things and invented ideas through which we live and that make up our identity. The culture that creates imagined communities is not only what is called high culture but, perhaps more importantly, is also culture in this wider sense.

This is because communal or national identity is not something that is laid over your personal identity, as if you were a blank canvas with a nationality painted on top. To ask which came first, the personal identity or the communal, is to ask a chicken-and-egg question. Our sense of communal identity plays a central role in constructing us, *and* it is something we ourselves construct. The critic and theorist Homi Bhabha argues that people are both the *objects* created by communal and national identity and the *subjects* who, in turn, create it. We are objects of it because we are *constructed* by our languages, histories, location and so on – our culture – but we are also subjects because we *act out* identity in all sorts of ways, usually cultural. We create it as it creates us. So Bhabha argues that identity is both *pedagogical* (taught to us at home, at school, in the community) and *performative*, performed, acted out and "done" by us in all sorts of ways. Obvious ways of performing your nationality or acting out a national culture might include supporting a national sporting team or being involved in a nationwide event, such as voting in an election or celebrating a national holiday like Independence Day or Thanksgiving. But there are other, smaller ways, too: following your local team for example. All these acts both define you and are examples of you defining yourself.

English as cultural heritage

This is why the idea of cultural heritage is important. The imagined community keeps what it values from its past: tangible things like historical sites, museum exhibits, battlefields and so on, as well as intangible things like stories, attitudes, ideas and beliefs. *Heritage* in this sense is made up of the cultural things that shape the "we" of the imagined community and, significantly, is a version of how the community wants to see itself. (The question this raises, of course, is who decides what is and isn't heritage and why. For example, why are some mansions or log cabins preserved and other forms of accommodation left to disappear?)

But central to English are not the places (like Walden Pond) or the material items of cultural heritage (like Emily Dickinson's family furniture). It is rather the intangible shared stories, attitudes and ideas that "everyone should know." It is these that make the subject English so crucial for ideas of national identity. Texts – like Shakespeare's plays, or Emily Dickinson's poems, or Emerson's ideas, for example – make up a reservoir of tales, ideas, images and values constructing and strengthening the idea of the imagined community, whether that is national or local or exists in some other way. However, as I have argued throughout this book, reading is as much about how we look as it is about what we look at. This shared agreement about values, messages or morals is not one that arises from having novels, plays or TV programs in common or even from admiring the same monuments. It comes from being taught to *interpret* them in the same way. This means that it isn't so much a shared knowledge of, for example, Shakespeare that makes up a literary heritage but rather the way in which we have been taught to *understand and interpret* his works. English as a subject teaches you a way to look at things. In the way it makes you produce papers and projects, it teaches (and, through assessment and grading, enforces) a way of making you act out this method of looking. English as a subject can be, and certainly was, a form of cultural heritage, aiming to help create a "we" by making us read and interpret in the same way. Because the subject is compulsory at school and is also highly regarded, it is a particularly strong way to bind people together.

This is one of the reasons people find the idea of "theory" quite threatening to culture and communal identity. If theory is, as I have argued, a range of different ways of looking at things, it means that the one way of interpreting literature is no longer unique. Theory is seen as a threat not just because it offers new interpretations of texts but also because it offers new ways of looking. New ways of interpreting don't construct the same "we" as before; in fact, they both teach and produce new forms of national identity.

Theory and multicultural heritages

Despite the power of the "we," there is *not really any one single culture* that everyone inhabits. A cultural identity is always the *result* of a binding together, and the "we" of that identity results from the interactions of lots of different cultures. With very rare exceptions (communities isolated by historical accident or through their own choices, for example), this has always been the case. However, importantly, the modern world is characterized by even more interaction among cultures than ever before; some people describe this as part of the process of globalization. We now (and perhaps always did) inhabit a *hybrid* society where different cultural traditions, ideas and assumptions try to rub together. They might all share a nation, but people brought up in different places, either within or outside the national boundaries, and people brought up in different classes or in different ethnicities or with different religions or expectations have, to a greater or lesser degree, different cultures. Although this is sometimes seen negatively as the cause of friction, multicultural mixing can also be a fantastic benefit. Salman Rushdie (b. 1947) is a novelist whose work explores this mixing of an array of cultures. He wrote that his work

> celebrates hybridity, impurity, intermingling, the transformations that come of new and unexpected combinations of human beings, cultures, ideas, politics, movies, songs ... Mélange, hotchpotch, a bit of this and a bit of that is *how newness enters the world.*

Rather than insisting on a single imaginary "culture," it is more accurate to discuss the mixing of an array of cultures in an ongoing conversation of cultures. And, if different cultures are mixing and conversing more and more, this means, most importantly, that different ways of understanding and thinking about texts must also emerge and mix.

The sense of "we" is changing. The study of literature and language is an opportunity to understand and to encourage an even more open multicultural society. As I suggested in chapter 6, the curriculum we study has not one canon but many different canons by different writers. However, these texts are often still studied in the same, traditional way. What needs to change is the way we look at texts – new ways of looking, new ways of studying English. This discipline, perhaps more than any other, with its strange mix of literature, language, identity and tradition, is a crucible in which new versions of identities are being formed and understood afresh.

English, literature and politics

However, this link between identity and literature means that the discipline of English has been a way in which interests have sought to shape people not just educationally but politically.

But what does politics mean in this context? Usually, when conversations turn to "politics," they tend to be about Democrats or Republicans, the most recent or upcoming elections or the personal qualities of people whose job it is to be politicians. But politics is really about much more than that: the word comes from the ancient Greek word *polis*, meaning "city," which hangs on in words like "metro*poli*tan," "metro*polis*" and – as characters in *Men at Arms* (1993) by the British comic writer Terry Pratchett (1948–2015) point out – "*poli*ce officer" and "*poli*tician." But it means much more than "city," also denoting "community" or, more widely, "society." Politics is about people, societies and how we live together, not just the events on Capitol Hill; the word covers an enormous area of human life. Of course, literature, too, is involved with people, societies and how we get along with one another. Dealing with the same issues in this way, literature and politics are inevitably bound together.

I have already shown how English was developed to mold people – a "political" process. That process is still functional today. Arguments that might appear to have a very limited relevance about what should or shouldn't be read, about the canon, about how people should talk and write or about assessment are actually arguments about how to form people's view of the

ENGLISH, IDENTITY AND POLITICS

world. Whether you think English as a subject is about personal growth, learning skills for the workplace or social world, understanding cultural heritage or offering cultural analysis, English is a very political subject, and all these things affect how we get along.

But English, as a subject, is perhaps even more closely involved with these debates about how we get along for the following reason: the study of literature is one of the very few subjects where individual experience is taken seriously as knowledge, as something you can learn about. Its hallmark is, as the critic Robert Young writes, the "value and attention it gives to the personal and the subjective." Other humanities subjects like history or geography do take account of what people say or write but only as evidence for historical or geographical work. The knowledge they develop is about, say, the development of a country or the use of land in a city. In English, the knowledge is about the representation, form and content of the person's experience. Robert Young goes on: while, for example, there are "plenty of histories of colonialism . . . [S]uch histories rarely considered the ways in which colonialism was experienced, or analysed, by those who suffered its effects." The same is true of all sorts of issues. Harriet Beecher Stowe's (1811–1896) novel *Uncle Tom's Cabin* (1852) is supposed to have profoundly changed opinions about slavery before the Civil War. More recently, work that focuses on and takes seriously women's experience, as represented in literary texts, has helped bring often hidden issues into the light and into discussions of how we might live together.

However, sometimes our ways of reading make it hard to see these texts for what they are. For example, regarding issues of race, one of the first steps in understanding the representation of the African-American experience has been a reevaluation of the canon. Just as the African critics I discussed in chapter 6 argued that some Europeans thought African writing was not literary but "anthropological . . . sociological, journalistic, topical ephemera," it was important to establish that much African-American literature was literary and was a valuable source for thinking about the experience of being black in America. Indeed, it was to this tradition of African-American writing that U.S. President Barack Obama (b. 1961) turned in trying to understand the complexities of race in America and his own "nightmare vision" of it. In his autobiography *Dreams from My Father*, he describes how he turned to the work of Langston Hughes (1902–1967), Richard Wright (1908–1960), James Baldwin (1924–1987), W. E. B. DuBois (1868–1963) and Malcolm X (1925–1965), whose autobiography was written in collaboration with Alex Haley (1921–1992), who wrote the best-selling novel *Roots* (1976). The work of the groundbreaking African-American critic Henry Louis Gates took this argument a stage further. Aware that the more traditional ways of interpreting literary texts were very "Western," that they had possibly questionable presuppositions, which meant that they might not grasp the importance or key issues in African-American literature, he and others have developed an approach more suited to dealing with texts from this tradition. Gates focuses on the idea of "signifying," which he sees as describing a way of reusing and recombining other texts to make a distinctive African-American way of seeing and engaging with the world and other people. As the experiences of African-Americans are expressed and are properly valued and understood, he argues, the awareness of the need for wider political and social change will grow.

Looking back over this book, much of it has been about the relationship between the *polis* and English. I outlined, for example, how the canon was a construction that reflected not just the need for a curriculum but also political motives and ideas, and I showed how Shakespeare has been used for political purposes. Rather than look at every aspect of this, because politics covers an enormous area, I shall examine the relationship between politics and the critical attitudes that influence ways of interpreting outlined in chapter 5.

Critical attitudes and politics

Although politics and literature are interwoven, how the interaction actually works is still an open question. One way of looking at this is to refer back to the broad critical attitudes I discussed in chapter 5. The *extrinsic attitude* moves from the text out to the context. It argues that literature is about the world and worth studying for what it tells us about the world. In contrast, the *intrinsic attitude* focuses on the text itself, its form and structure. It suggests that texts, especially "great texts," have an ingrained artistic value and so are worth studying in their own right. These two attitudes lead to very different understandings of the relationship between politics and literature.

The extrinsic attitude: Literature as politics?

Those who share the extrinsic attitude will have no problem explaining how literature is political. Because texts are about the world, they will also be about how we get along – that is, about politics. Some critics show, for example, how texts display ideas about the politics of the time they were written or the political ideas of the author. If the ideas of the author are not of interest for interpretation, the extrinsic attitude might suggest that the "voice of history" could be speaking through the text to reveal (without the author's knowledge) a range of taken-for-granted political ideas. Others, sharing this extrinsic attitude, will concentrate on how texts are used. One example of this is the cultural materialist approach to Shakespeare, looking at the way "Shakespeare" – both the plays and the institution – is a construct of present-day political, cultural and economic concerns. In this case, literature and ways of interpreting literature are seen as a political tool to be questioned, taken over or taken back. Where the political position of an approach or a text is hidden, the aim of the extrinsic critic is to uncover it.

Many of those who share the extrinsic attitude understand English to be "cultural politics." This is a rather catchall term for thinking about the relationship between politics and culture. Politics – how we get along – exists in many different cultural spheres. There are national, regional and local politics, for example. Politics is also involved in different spheres of what people do and where they are. You might hear about the politics of the workplace, say, or the politics of the playground. So culture, too, is political. As I suggested, culture gives us our sense of who we are and how we should be. It is all to do with politics. Cultural politics, then, is where politics and culture are interwoven. For example, the canon, a "cultural" idea, has political consequences. (Whose voices are we allowed to hear? What are they saying?) Conversely, a political idea, such as "everyone should be equal," has cultural effects. (Would a film that argued that people shouldn't be equal be acceptable? Would it be successful?) Cultural politics argues that politics is reflected in culture, and culture in turn reflects back and influences how we get along. If English is a version of cultural politics, then each text we study is a political event, and every text tries to convince us of certain ideas about how we should get along.

The intrinsic attitude: Literature versus politics?

In contrast, the intrinsic attitude implies a very different understanding of the interweaving between society and literature. For critics who share the intrinsic attitude, to see English as cultural politics is to miss the artistic worth, the "literary-ness," of a work of literature. To do English is to concentrate on the special features that make a work of literature great art. It is wrong, from this point of view, to look at the sociology, polemical messages or social intent of a literary text. In this sense, they might argue, literature is *counter* to politics and to the way in which people use power.

This idea has taken a great deal of criticism. It seems to imply that you could think about a text in a vacuum, separate from the world. It also implies that judgments about value can be unaffected by other opinions and ideas that you might have. For example, a work might offer a viewpoint about society with which you disagree completely, but you might still value it as a great work. Yet even this approach assumes that literature is involved in how we get along. Indeed, if the power of literature lies in the ways in which experience is represented and structured, perhaps it is the "great works of art" that do this most acutely. Some people who follow the intrinsic attitude argue that literature teaches moral truths or that it embodies the human spirit. But you don't have to believe this to see that an artwork can have an amazing, unpredictable and transformative effect on people's lives and situations. Exploring and perhaps fostering these effects is, in the end, about how we get along, which is why studying literature is about politics in the widest sense.

Why has English been a political battleground?

The question remains, "Why do politicians, teachers, academics, journalists, bloggers and others argue so much over the subject of English?" There are a number of reasons. Perhaps most importantly, ideas clash over English because, as such a popular subject, it is one of the larger forums in which many people encounter a structured approach to cultural issues. As I have suggested, cultural activity, especially education, plays a large role in shaping and controlling our ideas about such things as identity and social hierarchies, and it has a huge influence over our worldviews. Studying literature, as a part of culture in general, is a very powerful way of forming people, not least because it is there, most often and most clearly, that people's experience is represented. So when people seek to shape ideas, to convince others and to make changes in society, the subject of English is one of the tools to which they turn. So-called identity politics is a very powerful force, and literature is often used in these discussions. But literature and the study of literature can also be used to question ideas and assertions about identity, extend sympathy and raise questions. It can – perhaps should – make us question and examine our presuppositions rather than simply affirm them.

This process of shaping and molding has become even more important because we live in an age of mass and social media communication, where the way we represent things has become much more significant. Think of the debate over the canon discussed in chapter 6, for example. Including works by those who have traditionally been considered a minority (texts by African-American women, say) is a form of representation. Studying a wider range of texts offers a broader and more open representation of the world, and, as a result, it also might help to prevent people from being excluded not only when we think about what literature might be but also when we think about what society might be. Those with an interest in such issues have seized upon English as a forum for discussing their ideas.

English is also controversial because it is in many respects one of the most important subjects in education. This is not because knowledge about Shakespeare, for example, is more important than being able to do math (because it isn't), but more simply because English focuses on issues of interpretation and subjective knowledge that other subjects often pass over. This key role means that as people – politicians, teachers, academics or the media – try to influence the education system as a whole, they often turn first to English. As a result, English becomes both a test tube for education policy and a weathervane, showing which ideas are strongest at any time in education as a whole. This could explain why those interested in education react very strongly to any proposed changes to English courses.

Studying English, then, makes us sensitive to how we get along, to the *polis*. To do English is to become involved with others, through literature and language. It leads you to uncover ideas other than your own and new ways of thinking about things. We might think of reading as a private, solitary activity, but all the time it is forming links between you and others in the world. Reading by yourself is, in fact, one of the most social, political activities you can do.

Summary

- English is intimately involved with questions about identity because identity is constructed culturally.
- Communities exist *in the imagination*, built upon by a shared stock of images, ideas, stories and traditions. We are all both the *objects* created by cultural identity and the *subjects* who, in turn, create it.
- Those cultural things from the past that are chosen to shape the "we" of the imagined community make up a cultural heritage. Traditionally, English, the subject, is a form of cultural heritage, both in the texts it chooses (the canon) and in the way it interprets those texts.

- Politics can be defined in its broadest sense as how we get along. English, in dealing with literature, also deals with ideas about society and our place in the world. The two are inextricably linked, not least because of the way literature deals with the individual's experience.
- Those who support *extrinsic* forms of criticism suggest that texts are about the world and that English is a form of cultural politics – a point where politics and culture are interwoven. Those who support *intrinsic* criticism would disagree with this but still acknowledge the link between literature and how we see our place in the world. Whether these critics realize it or not, this makes English a political activity.
- English is a site of controversy because it is an inherently political subject. Issues of representation within English courses are increasingly seen as important in the wider world, so the subject is a focus for those interested in such issues. In addition to this, the interpretative skills taught in English are at the base of all other subjects, so anyone wishing to change education must engage with English.

13

Why study English?

- Why major in English?
- Is English useful?
- Is English valuable?
- What skills does it involve?

The quick answer to the first question is because it's great, plus skills. I wouldn't be writing this book if I thought it wasn't worthwhile. But as I've said throughout the book, everything about English is contentious, even – perhaps most especially – the question of why someone might choose to study it. It's a question that deserves an answer. Students give all sorts of good reasons for wanting to major in English: because reading is pleasurable and fun; because it's interesting; because it allows them to experience a huge range of thoughts and feelings; because they like the different ways it can be taught, through conversation, which leads to forms of self-discovery; because works of literature speak to them. All these reasons combine to make many English majors noticeably more enthusiastic about their study. This is especially gratifying to teachers and professors of English because there is supposed to be a so-called crisis of the humanities. Many of the traditional arts and humanities subjects, centrally English, seem to be in decline and under attack: where once they appeared so central, now they look as if they are being pushed to one side by economic or other necessities. In contrast to science, technology, engineering and math, the STEM subjects, what use is English? What skills does it teach? But this is English: of course, even the question of "use" is controversial.

What's the use of "use"?

Should we think of English as useful or not? Often this debate is cast in terms of thinking of education as "instrumental," that is, as a tool for doing something. It is also discussed in relation to utilitarianism. This philosophy, most clearly expounded by British thinker Jeremy Bentham (1748–1832), argues that what matters is only how useful something is. For its use in making people happy, Bentham wrote, "the game of push-pin [a child's or bar game involving pushing pins – the equivalent of Tetris or Angry Birds or this month's simple but gripping game app] is of equal value with the arts and sciences of music and poetry." For people who are interested only in how useful something is or how it can make money, English at first doesn't look very beneficial for the individual, the economy, the community or the nation.

In contrast to this, some argue that literature and arts are special, and even though, of course, literature is used in some ways (Shakespeare made a living from his plays), there's something about it that is simply outside use value – a literary text isn't simply a tool for doing something else but is innately significant in itself. Moreover, to think of it instrumentally is to betray precisely the beauty, passion, interest, love and self-understanding that students (and teachers!) feel in literature: we don't measure the people we love by their use, after all. The study of literature, of what the Irish poet William Butler Yeats (1865–1939) called "monuments of unaging intellect," is simply a good and important activity in its own right.

Further, some argue that to reduce literature to its use is to turn literature and its study simply into another sort of commodity, like a cup of coffee or pair of jeans, which can be traded and exchanged for cash or kudos (so-called cultural capital): that, too, betrays what it is. This idea also implies that literary understanding can simply be bought and doesn't have to be learned through a process of reading, thought and reflection. (As a comparison, rich people might be able to afford the best clothes or coffee easily, but, although they can buy the time to exercise, to get fit they still have to sweat and strain like everybody else: perhaps that's a closer analogy to coming to understand literary texts.)

Thinking of education as a tool can also have some odd side effects. "To a man with a hammer," the old saying goes, "everything looks like a nail":

academic enquiry is supposed to be disinterested. This doesn't mean boring but rather not serving any particular interests: a lawyer, for example, might read a case report with the interests of a client foremost in mind. If one reads a text with too limited a range of questions in one's mind, like the hotel guest reading *Macbeth* in chapter 6, texts can begin to look limited. To read a text just for what's on the test is to misunderstand something vital (and would miss, ironically, what's crucial for doing well in the test: reading and thinking as a critic).

One reason for this feeling that "usefulness" is a betrayal can be explained historically. As Louis Menand explains, Charles William Eliot (1834–1926), who became president of Harvard in 1869, helped shape U.S. universities. Importantly, he played a central role in developing both the elective system for undergraduates and professional graduate schools. For the undergraduate college, Eliot wrote in 1869, the "dominant ideas" should be "the desire for the broadest culture, for the best formation and information of the mind, the enthusiastic study of subjects for the love of them without any ulterior objects, the love of learning and research for their own sake." In contrast, in the more vocational schools (in engineering or law or medicine, for example), the student is training "with the express object of making himself a better manufacturer, engineer, or teacher; he is studying the processes of nature, in order afterwards to turn them to human uses and his own profit; if he is eager to penetrate the mysteries of electricity, it is largely because he wants to understand telegraphs." Eliot continues: the "practical spirit and the literary or scholastic spirit are both good, but they are incompatible. If commingled, they are both spoiled." Usefulness is seen almost as a cause of rot for literary study.

The values of English

Others, however, argue not against the usefulness of the humanities and the study of literature but for a wider sense of their *value*. Helen Small sums up some of these views. She says that, in the face of demands simply to be useful or to fall in with what "we all need," the study of the humanities, including literature, remains a valuable place in which individuality and personal distinctiveness can flourish. Moreover, this sort of study "assists in the preservation and curation of the culture, and of the skills for interpreting and reinterpreting that culture to meet the needs and interests of the present." The humanities also make a "vital contribution to individual happiness and to the happiness of large groups" in ways that are not simply economic or instrumental.

The philosopher Martha Nussbaum takes some of these arguments further. The study of humanities subjects, she argues, is vital for democracy.

> Democracies have great rational and imaginative powers. They are also prone to some serious flaws in reasoning, to parochialism, haste, sloppiness, selfishness, narrowness of spirit. Education, based mainly on profitability in the global market magnifies these deficiencies, producing a greedy obtuseness and a technically trained docility that threaten the very life of democracy itself, and that certainly impede the creation of a decent world culture.

For Nussbaum, the humanities are vital to make democracy function properly and to live up to its promise. Citizens need not just factual knowledge but the more complex ability to assess and weigh up that knowledge; citizens need to follow, understand, judge and make arguments – to think critically; citizens need to have the fullest sense of the importance of each of their fellows across the world; and citizens need to have what she calls "narrative imagination" – the ability to see sympathetically and flexibly, as far as possible, from other people's point of view, from within others stories of themselves. This is good for business too: she writes that leading "business educators have long understood that a developed capacity to imagine is a keystone of a healthy business culture. Innovation requires minds that are flexible, open and creative; literature and the arts cultivate those capacities." (Successful businesspeople will tell you that you don't know where the next good idea is going to come from. Perhaps the most famous English professor in the world is J. R. R. Tolkien (1892–1973) who studied and taught the relatively obscure area of Anglo-Saxon verse. His stories – centrally *The Lord of the Rings* and *The Hobbit*, informed and shaped by that research – have made billions of dollars for the book, film and gaming industries and have created many thousands of jobs). For Nussbaum, it is through the humanities that the crucial virtues of democracy are most effectively learned. However, as I suggested looking at the history of the discipline of English, questions of value are extremely contentious.

All these views are important. By thinking about *how we read* and the presuppositions that reading involves, it's clear that each of them offers implicit and different ideas about what literature is and what it does. All these views are important. By thinking about *how we read* and the presuppositions that reading involves, it's clear that each approach offers implicit and different ideas about what literature is and what it does. But sometimes it sounds as if there is a divide between the noble aims imputed to literature and the experience of reading a great work, on the one hand, and, importantly in this context, the mundane-seeming academic study of literature, on the other. Students of English love literature, and they also (usually!) want to do well in tests and learn useful skills. And I don't think that, if "commingled," both these aims are "spoiled." In fact, education at its best is a holistic process in which the different parts are connected in the light of a whole. And, as Menand again argues, that "knowledge just *is* instrumental: it puts us into a different relationship with the world." He goes on to make the old joke that "Garbage is garbage but the history of garbage is scholarship" and adds, "Accounting is a trade but the history of accounting is a subject of disinterested inquiry – a liberal art. And the accountant who knows something about the history of accounting will be a better accountant. That knowledge pays off in the marketplace."

But why should knowing about the history and philosophy of accountancy – rather than just knowing the rules and procedures – make one a better accountant? The main reason is that, like every complex human activity, accountancy is constantly changing and developing, as new ideas, practices and technologies, as well as theoretical arguments about its nature, shape how it's done and its relationship to business and society as a whole. Understanding the principles of accounting, along with some of the struggles and controversies that have made up its history, means one is better prepared to deal with new situations and complex change, as well as understanding its evolving role. A good accountant has learned the disciplinary consciousness of accountancy. Accountants learn it by reviewing case studies and following financial and legal discussions and protocols; English students learn theirs (I'm pretty sure more delightfully) by reading and discussing novels, poems, plays, other texts and works of criticism and theory. Importantly, learning one disciplinary consciousness makes it much easier to learn another: this is crucial in learning how to learn. And perhaps in English, because it is such a self-reflexive discipline where the controversies mean that the disciplinary consciousness is continually brought to the fore, this process can be made very clear.

"It taught me skills?"

Ideally these two strands of education – its values and its instrumental and practical use – run holistically together, as I've said. So, having discussed values, I want to turn to some of the skills that English as a discipline teaches. For many in English, even to talk of skills is to betray the subject, as I have suggested, and it does feel uncongenial to discuss the experience of reading in this way. But in a discussion of studying English, it's important. A student of mine was applying for a graduate job in a major publishing firm and asked me to look over her resume. It mentioned her degree in only one line but spent a page listing the skills she had learned during her demanding part-time job working in a bar. I asked her if the publisher would be more interested in those skills or the skills she had learned from her degree in English. "It taught me skills?" she asked, looking at me blankly. She wasn't to blame: we were, for not making plain the skills she was learning. English as a subject has always been bad at discussing the business end of what it teaches and its link to employment. Perhaps it is too keen to stress the other benefits – moral, communal, intellectual – of an education in the humanities or the enjoyment found in reading and discussing works of literature. But English as a subject does teach real skills and can afford to be more explicit about them. In an article called "Fear of Being Useful" (2012), Paul Jay and Gerald Graff quote the CEO of State Farm Insurance:

> [O]ur employment exam does not test applicants on their knowledge of finance or the insurance business, but it does require them to demonstrate critical thinking skills . . . the ability to read for information, to communicate and write effectively, and to have an understanding of global integration.

Generally, and as information becomes more and more easily accessible, employers are valuing the critical thinking, imagination and creative flexibility of English graduates.

However important skills are, lists of skills are tendentious (one asks, "Do I have all of them? How good am I at them?"). They can also be boring to read and can encourage a check box attitude. However, in part to put these on record, I am going to provide a list here. One might add, a bit tongue-in-cheek, that the people who don't mind constructing lists of boxes to check are bureaucrats. And sure enough, the British government's Quality Assurance Agency produced a document, written for the most part by professors of English, that recorded among other things what sorts of skills and knowledge an English graduate should have. I quote here liberally from that document – you can view it in full online – but more as an illustration of the sorts of skills than a definitive list.

First, there is the *subject knowledge* you learn from English. This involves knowing about a range of literature and ideas – an awareness of how culture, language, technology, economics, gender, race, body and many other factors can affect how, where and by whom texts are produced, received, understood and appreciated. It also involves understanding the role of readers in shaping texts; the relationships between different genres and different media; critical, theoretical, linguistic and stylistic concepts and terminology; form and its importance; and, through following works of literature, understanding and analyzing debates, arguments and ideas over time.

There are also *skills specific to English*. People who have studied English can read closely and critically, paying attention to the details of texts. They can analyze texts and discourses, respond to the affective power of language, using appropriate approaches and terminology. Crucially, they develop independent and imaginative interpretations of literary, critical, linguistic or creative material and can articulate a critical understanding of complex texts and ideas (and of their historical relations). They can write clearly, accurately and effectively and apply scholarly bibliographic skills appropriate to the subject.

There are also *generic and graduate skills*. English graduates are effective researchers, good communicators and active learners. As *effective researchers*, they can discover and synthesize complex information and diverse evidence and respond creatively and imaginatively to research tasks. They can initiate projects of their own. English graduates can present information within wider contexts; test, interpret and analyze information and evidence independently and critically, producing from that analysis cogent arguments and decisive judgments and plan; and organize and report by the deadline. As *good communicators*, they possess advanced communication skills and are able to articulate their own and other people's ideas concisely, accurately and persuasively, both orally and in writing. They can develop working relationships with others in teams, especially through constructive dialogue (for example, by listening, asking and responding to questions). They understand the role of narrative and emotion in decision making and can be sensitive to cultural contexts when working with others. And as *active, lifelong learners*, people who have studied English can adapt to different demands and tasks, appreciate the benefit of giving and receiving feedback, evaluate and reflect on their own practices and assumptions, look beyond the immediate task to the wider context, including the social and commercial effects of their work, and initiate and take responsibility for their own work. In the information-rich age, learning how to learn is central, and English, with its insistence on self-reflection, is a superb discipline to undertake that.

Put like this, the skills seem rather overwhelming, but, of course, these skills are holistic and are learned together. Athletes train in several different ways in different sessions because they know that different forms of exercise build different strengths and skills together. They don't, for example, spend weeks just building up their upper body strength while ignoring their stamina and legs. Moreover, learning is a *cumulative process*: you are involved in the *process* of learning these skills, and nobody expects you to have them all immediately. Ideally, these skills are *embedded* in the course you are studying. When you write a paper, for example, you are improving your *subject knowledge* of a period or writer, understanding how that text, say, fits into a history or aesthetic movement or debate. You are using and improving your critical, theoretical, linguistic and stylistic concepts and use of terminology. And you are also working on your subject skills: reading closely; analyzing and responding to text; writing clearly, accurately and effectively; and discovering what complex ideas and debates are like. Writing a paper means uncovering information and bringing it together, coming to develop your own opinions and ideas, following up on things that you have found interesting, thinking things over and weighing them up, and then explaining what it is you think and why. It also means taking feedback and responding to it to improve your future performance.

In class and in seminars, you develop knowledge by listening and discussing texts and ideas in their historical, cultural and intellectual contexts, looking at the creative, cultural and intellectual forces active in shaping them, their reception, the contemporary debates they may be involved in, their use of language, their form and their genre. This focuses your attention on issues and texts, providing techniques for analyses and heightening your sensitivity to language and contexts. It will help you assess the relevance of your research and ideas, use evidence, make constructive comparisons and contrasts, understand the place of both detail on the one hand and overarching structures and designs on the other and judge the appropriate level of detail. It will also develop your understanding of the terms, procedures, process, content and style of others' arguments, which in turn leads to the effective use of the terms, procedures, process, content and style of argument yourself. Seminars will help you draw conclusions, follow both lateral and logical inferences, evaluate and judge arguments and question your own and others' assumptions and presuppositions. Finally, seminars develop your communication skills through presenting prepared work to peers and assessors, through thinking on your feet by responding to questions, through following discussions and engaging in dialogue, group discussions and group reporting and through questioning peers and staff. Through learning how to respond effectively to tasks, questioning peers and staff, reflecting on your own practices, thinking independently, participating in teamwork and cooperation situations and developing self-discipline, both intellectual and social, you also learn how to learn.

Conclusion

As I suggested at the start of this chapter, there is a great deal of discussion about the crisis in the humanities. However, as Michael Bérubé argues, the numbers of undergraduate enrollments are not in decline:

> [D]espite skyrocketing tuition rates and the rise of the predatory student-loan industry . . . despite decades' worth of rhetoric about how either (a) fields like art history and literature are elite, niche-market affairs that will render students unemployable; or (b) students are abandoning the humanities because they are callow, market-driven careerists, *despite all of that*, undergraduate enrollments in the humanities have held steady since 1980 (in relation to all degree holders, and in relation to the larger age cohort), and undergraduate enrollments in the arts and humanities combined are almost precisely where they were in 1970.

He rightly points out that arguments over a decline (or not) in English is itself the site of controversy. Those who claim there's been a decline, he argues, are really attacking current ideas in the study of English like the "study of race, class, gender . . . the rise of 'theory'. . . or the study of popular culture . . . or the fragmentation of the curriculum." It may be true that not everything in the humanities or in English is going well, but the enthusiasm, passion and commitment of English majors seem to be surviving.

Philip Sidney (1554–1586) argued that the work of the poet, more than the historian or moral philosopher, educates "the mind more effectually than any other art doth" because it teaches with the "hand of delight": it's great, plus skills. It might be right to stress the anti-instrumental side to studying English, how the discipline questions use and how it champions values; even doing this helps foster vital critical skills that empower students. These two sides can't really be separated from each other because they are part of the holistic experience of studying. And, if English majors and their professors do look to more vocational disciplines in envy or in pity, thinking them imprisoned by the need to be useful, do we think that those majoring in STEM subjects don't ever wonder, in a far from instrumental way, at the beauty of the universe, the intricacies of materials or the mysteries of mathematics?

Summary

- In the arts and humanities and in English, the question of use is controversial. Some argue that to look at the subject focusing on its benefits to the economy or the state is mistaken and that the study of literature is simply a good thing in its own right.
- Others stress the values of English: it helps foster individuality and preserves cultures.
- It may also play a key role in a democracy, in teaching citizens to assess information, to make and judge arguments, to think critically and understand one another better and with more sympathy.
- The study of literature also teaches skills, and its graduates are effective researchers, good communicators and active learners.
- These skills are ideally holistically embedded in courses you take.

Conclusion

The importance of English

This book is about *why* and *how* we do English. It is a book about ideas and has explained what you are doing when you are doing English. The book has covered some key ideas for English today.

Many of the ideas I have described affect your assessment, grades, papers and choice of texts, aims, objectives and everything else to do with English, usually without your knowing. I think it's important to see how and why things are done the way they are. (Ben Knights and Chris Thurgar-Dawson write suggestively that the "history of a discipline is, among other things, a covert history of the relation between the culture of the staff tribe and that of the student tribe": part of the point of this book has been to uncover this secret history, to look "backstage.") This is not least because, if you know why you are doing something, it makes it much more straightforward to do it. As you progress in English, you will realize that I have simplified ideas and issues from time to time: because it's often assumed that everything you do in English should be naturally accessible, simplification is often frowned upon. But again, this book is a tool or, to use a famous metaphor from Ludwig Wittgenstein, a ladder to be thrown away after use. (You might begin the process of throwing it away by thinking about what's wrong with the model of reading presented in chapter 3.) I have no control over how anything in this book might be interpreted: what you make of it is up to you. But isn't that the case with every text?

I have suggested that English is rapidly changing. It used to be the case that in order to succeed in the study of literature, students had to learn to look through one set of eyes, perhaps very different from their own. In doing so, they accepted, often without realizing it, the worldview behind the discipline that was developed as a subject in the first half of the twentieth century. Among other things, this turned potentially exciting literature into bland exam fodder. All this risked making English into a "heritage" subject that was studied as an insipid ritual and that ran roughshod over the fact that students at all levels come from different backgrounds, have different formative experiences and bring different presuppositions.

The situation has changed, however, and one sign of this change is the wider diffusion of new ideas and innovative ways of looking at literature, often through what is called theory. Theory isn't just an arbitrary collection of names and jargon. Theory – as I've said, not a good term – stands for the new questions that readers ask of literary texts that weren't asked before (and so it reveals that older approaches have particular questions that they ask and issues that they pursue). We ask new questions because the world has changed and the need for new ways of thinking about it stems from that change. For example, we read feminist theory and criticism not to play at being feminists; we read them because there are important questions to be asked about women's roles and lives and about gender issues. We explore the relationship between literature and the environment because literature helps us focus on and understand our relationships with the ecosystems we inhabit. And we discuss global literature not to pose as cosmopolitans but because we all inhabit a globalized world and because how we understand and live with the consequences of this is crucial, as events all over the world clearly show. English – including theory – is not an abstract glass bead game played only for pleasure; it is a discipline that, through the study of literature, attempts to comprehend better the world around us and to appreciate the others who inhabit it. How we read is part of how we are in the world we share, and questions about this arise from reading texts. Our reading and interpretation grows out of our experiences, concerns and hopes for ourselves and others.

This book, then, has sought to look at some ideas that shape the study of literature today and in this way to answer the "why are we doing this" when we study English.

To summarize:

- Reading is an active interpretation, and English deals with texts and how we read. Once we are aware of different ways of interpreting texts, it becomes clear that there is *no neutral, objective approach* to literature. In turn, this means that there could be no single method of studying English, new or

traditional, and no single correct interpretation. I feel that we should watch out for replacing "one way" simply with "another way." English is a pluralist subject (it accepts a wide range of approaches) and is open-ended.
- Like all disciplines, English has a disciplinary consciousness; to study English means to learn to think *as a* literary critic. This involves not only knowing about literary and other texts but also understanding debates and ideas about what literature is, different forms of interpretation, literary value, the canon, authorship, the uses of language, narrative and other matters. English has a complex history that still shapes its current practices.
- English, as culture and as a subject that studies culture, is *involved with our relationships with others and with the world*. Culture is woven inextricably into how we get along and has far-reaching effects in the wider world. A consequence of this is that English is not just about texts but also about you, about others and about the nature of society. It also means English is a very controversial subject.
- English, like all disciplines, is constantly evolving and changing, drawing on new ideas, following internal debates to new conclusions and reflecting society.

None of this is to argue that "anything goes" in English. Looking at texts, interpretation and a wide range of significant ideas, then relating this to our cultures and societies, involves knowledge and careful thought. Perhaps most of all, it involves constantly taking *responsibility* for each interpretation. English also asks, "*Why* do you think that about the text?"

Nearly all education has two strands to it. One is about learning skills and facts for their own sake, proving you have them by passing exams and getting a qualification, which in turn opens doors. The other side to education seems less concrete but is, in a way, closer to each of us: it is about fulfilling your own potential, following your own interests, exploring yourself and others, your society, the world, and becoming – often in some indefinable way – better. But just because these aims sound vague and you couldn't be assessed on them does not mean that they are not valuable. In fact, some might say that these are truly the point of education. Nearly forty years ago, the educationalist Harold Rosen (1919–2008) stressed precisely this side of the subject of English:

> [It is] nothing less than a different model of education: knowledge to be made, not given; knowledge comprising more than can be discursively stated; learning as a diverse range of processes, including affective ones;

educational processes to be embarked on with outcomes unpredictable; students' perceptions, experiences, imaginings and unsystematically acquired knowledge admitted as legitimate curricular content.

This is stirring and right: I drew on these ideas explicitly in chapter 1. But there are exams, too. Sometimes the assessment and fulfillment sides of education run together, and the whole process can be a joy. Sometimes, they don't, and getting the "right answer" kills off some of the very real interest. Theory, with its concern to offer different ideas and different voices and to validate different approaches, is, among other things, part of an attempt to bring these two sides together.

Moreover, while the so-called moral mission of English turned out to be an illusion – subtly coercing people to share the views of a certain type of person of a particular class, color, sex and certain age – there is a link between the study of literature and ethical responsibility. Many people argue that the all-encompassing ideas and systems that led people to take some central beliefs for granted have collapsed or are in the process of collapsing. For example, even if we choose a system by which to orientate ourselves – a set of political beliefs, religious beliefs or a philosophical approach to the world – we have usually actively chosen it rather than just simply accepted it, as people in the past might have done. The result of this is that each of us is more *responsible* in two ways. First, decisions, especially decisions about doing the right thing, have to be argued and negotiated, even though there may be no absolute surefire way of proving them "correct." The burden of this now falls on each of us, not on a system of beliefs to which we adhere. Second, and because of this, we have to be sensitive enough to *respond* to each situation and each choice as best we can. This involves not just viewing the situation as fully as we are able but also reflecting on the ideas and approaches that led to that particular interpretation. And English as a subject has a role to play here, in making us more reflective and responsive.

Some people argue that literature shows us other people's experience or that it teaches us to "walk a mile in somebody else's shoes"; I mentioned one example in the last chapter, Martha Nussbaum's idea of narrative imagination. This experience, they suggest, makes us more responsive to other people's needs, ideas, hopes and fears. The trouble with this idea is that, even after reading a book or poem that does this (and, of course, not all do or are interpreted as doing so), it is still possible to forget or to assume that this one story is only a story. Walking a mile in somebody's shoes is walking only a mile, and a sensitivity can soon become callused again. Our responsiveness is perhaps better developed by thinking about *how* we read. By understanding different presuppositions and by uncovering what we take for granted, it is possible to develop a habit of constantly questioning whatever you read or see or think or do. This constant questioning in turn develops a heightened responsiveness.

You, as a reader and student of English, should be free to explore many methods of interpretation, or to hop from one to the other, or to experiment with a selection. By consciously seeking out and using different methods of interpretation, motivated by presuppositions different from our own, each of us can bring to light, learn about and, perhaps, challenge our own preconceived ideas. This leads to ideas about works of literature that are new, interesting and exciting in themselves but also helps us to see the world differently. In this way the power of literature is clear: it can continue to unsettle us and to make us question even our most closely held beliefs, not only about art but also about ourselves, others, society and the wider world.

The philosopher Hannah Arendt (1906–1975) celebrated what she called "natality," the interruption into life of birth. She goes on:

> [The] new beginning inherent in birth can make itself felt in the world only because the newcomer possesses the capacity of beginning something anew, that is, of acting. In this sense of initiative, an element of action, and therefore of natality, is inherent in all human activities.

Indeed, it is this possibility of unpredictable newness, fully experienced, that bestows "faith and hope upon human affairs." And it is this sense of "unpredictable newness," really, in that part of "human affairs" that involves literature, its reading, writing and interpretation, that this book has been about. English is a contested and constantly evolving discipline. New ideas, new books, poems and plays, new ways of reading and writing and the new demands these make create its unpredictable newness. No one knows what will happen next, and we have to strive to be open to what we can't yet know. This means that the future – of the subject, of how we understand literature, of literary creation – lies with you, the active reader, now, studying English.

Further reading

They say that students, teachers and academics are just the reproductive system of libraries – after all, each one recreates a little library. If this is true, then the DNA of libraries is encoded in bibliographies. A bibliography serves two purposes: to show where the ideas you have been reading about came from and to provide a list of further things to read. This bibliography aims mainly to serve the second of these purposes (though all the critical works I have cited are mentioned here). It offers a first port of call, so to speak, for what to read next.

1: Studying English

It's hard to find a book that sums up English. David Foster Wallace (1962–2008), in a piece widely available online, puts it like this:

> "Critical appreciation" means having smart, sophisticated reasons for liking whatever literature you like, and being able to articulate those reasons for other people, especially in writing. Vital for critical appreciation is the ability to "interpret" a piece of literature, which basically means coming up with a cogent, interesting account of what a piece of lit means, what it's trying to do to/for the reader, what technical choices the author's made in order to achieve the effects she wants and so on. As you can probably anticipate, the whole thing gets very complicated and abstract and hard, which is one reason why entire college departments are devoted to studying and interpreting literature.

The book by John Hattie is *Visible Learning for Teachers: Maximizing Impact on Learning* (London: Routledge, 2011).

2: Where did English come from?

There is a growing number of books on the history and origins of English as a subject. For this chapter, these were especially useful:

- Michael Gardiner, *The Constitution of English Literature: The State, the Nation and the Canon* (London: Bloomsbury, 2013).
- Gerald Graff, *Professing Literature: An Institutional History* (Chicago: Chicago University Press, 2007).
- Gerald Graff and Michael Warner (eds.), *The Origins of Literary Studies in America* (London: Routledge, 1989).
- Geoffrey Galt Harpham, *The Humanities and the Dream of America* (Chicago: University of Chicago Press, 2011).
- Louis Menard, *The Marketplace of Ideas* (New York: W. W. Norton, 2010).
- Francis Oakley, *Community of Learning: The America College and the Liberal Arts Tradition* (New York: Oxford University Press, 1992).
- James Turner, *Philology: The Forgotten Origins of the Modern Humanities* (Princeton, NJ: Princeton University Press, 2014).
- Ted Underwood, *Why Literary Periods Mattered: Historical Contrast and the Prestige of English Studies* (Stanford, CA: Stanford University Press, 2013).
- Gauri Viswanathan, *Masks of Conquest: Literary Study and British Rule in India* (New York: Columbia University Press, 1989).

Other important studies are:

- Carol Atherton, *Defining Literary Criticism* (Basingstoke: Macmillan, 2005).
- Chris Baldick, *The Social Mission of English Criticism, 1848–1932* (Oxford: Clarendon Press, 1983).
- Brian Doyle, *English and Englishness* (London: Routledge, 1989).

Roger Geiger's *A History of American Higher Education: Learning and Culture from the Founding to World War II* (Princeton, NJ: Princeton University Press, 2015) provides a fuller historical context to much of this discussion.

John William's novel is *Stoner* (London: Vintage, 2012), and the Sherlock Holmes story I refer to is "The Adventure of the Golden Pince-Nez."

3: Studying English today

There are a number of introductions to literary theory and English today These are among the best:

- Lois Tyson, *Critical Theory Today: A User-Friendly Guide* (New York: Routledge, 2014). This is just what it says! Highly recommended.
- Andrew Bennet and Nicholas Royle's *This Thing Called Literature: Reading, Thinking, Writing Paperback* (New York: Routledge, 2014) is a very interesting, innovative introduction to key critical concepts.
- Peter Barry, *Beginning Theory: An Introduction to Literary and Cultural Theory* (Manchester: Manchester University Press, 2009).

Vincent Leitch's *Literary Criticism in the 21st Century* (New York: Bloomsbury, 2014) is not a guide but does introduce much in the current state of the discipline.

There are two large anthologies of theory:

- Vincent B. Leitch, William E. Cain, Laurie A. Finke, Barbara E. Johnson, John McGowan, T. Denean Sharpley-Whiting, and Jeffrey J. Williams, *The Norton Anthology of Theory and Criticism* (2nd ed.) (New York: W. W. Norton, 2010).
- Julie Rivkin and Michael Ryan (eds.), *Literary Theory: An Anthology* (Oxford: Blackwell, 2009).

And a useful glossary:

- M. H. Abrams and Geoffrey Harpham, *A Glossary of Literary Terms* (9th ed.) (Boston: Wadsworth, 2009).

Other introductions are:

- David Ayres, *Literary Theory: A Reintroduction* (Oxford: Blackwell, 2008).
- Hans Bertens, *Literary Theory: The Basics* (London: Routledge, 2001).
- Jonathan Culler, *Literary Theory: A Very Short Introduction* (Oxford: Oxford University Press, 1997).
- Terry Eagleton, *Literary Theory* (3rd ed.). (Oxford: Blackwell, 2008). The first edition of this best-selling introduction to theory was published in 1983, and, although a new edition was produced in 2008, there were almost no changes, so it now looks very dated.
- M.A.R. Habib, *Modern Literary Criticism and Theory: A History* (Oxford: Blackwell, 2008).

Ludwig Wittgenstein's metaphor is from *Philosophical Investigations* (Oxford: Blackwell, 1963).

FURTHER READING

4: English and disciplinary consciousness

English is one of the most discussed subjects in the curriculum, and many books discuss its pedagogy. Sections of this chapter draw heavily upon Patrick Scott's excellent (but now a bit dated) book about English in the UK, *Reconstructing A-Level English* (Buckingham: Open University Press, 1989).

This chapter is also influenced by the work of Ben Knights, including his "Intelligence and Interrogation: The Identity of the English Student," *Arts and Humanities in Higher Education* 4 (1) (2005): 33–52. Ronan McDonald, *The Death of the Critic* (London: Continuum, 2008) is also worth reading on these matters.

The brief discussion of tradition is influenced by Alastair MacIntrye, *After Virtue* (2nd ed.) (London: Duckworth, 1985), pp. 221–222. In his book *Organizing Enlightenment: Information Overload and the Invention of the Modern Research University* (Baltimore, MD: Johns Hopkins University Press, 2015), Chad Wellmon discusses the ways disciplines develop. In an interview with Scott Jaschik (in *Inside Higher Ed*, May 8, 2015), he writes that "knowledge was not some ready-made thing that dutiful scholars simply collected and displayed in printed tomes of erudition." He goes on:

> Knowledge was a problem that could never fully be solved. As opposed to mere "facts" . . . knowledge had to be crafted, shared and cultivated over time. It required research, not simply erudition. . . . This ethic came to be embodied in the ideal of what I term the disciplinary self and its distinct virtues: industriousness, attention to detail, a critical disposition and a commitment to the collaborative development of knowledge.

This means, he argues, that

> disciplines aren't just abstract taxonomies of knowledge. They are formative practices. Disciplines don't just organize ideas and concepts; they form particular types of people. They entail institutional and social practices and an underlying ethos aimed at producing a certain type of person who can then produce a certain type of knowledge . . . [D]isciplines are embodied as opposed to simply theoretical forms of knowledge. If we think of disciplines as traditions, then we're immediately dealing with questions of authority, transmission, language, concepts and, especially, critique. Traditions are never univocal or

monolithic. Disciplines are defined by perpetual internal conflict about what belongs and what doesn't, what is relevant for today and the future. The boundaries and standards of disciplines are highly flexible and are constantly being redrawn and rearticulated.

This is another way of thinking through disciplinary consciousness.

5: Critical attitudes

René Wellek and Austin Warren's ideas about intrinsic and extrinsic are developed in *Theory of Literature* (3rd ed.) (Harmondsworth: Peregrine, 1963), one of the most famous New Critical discussions of what literature is.

6: Literature, value and the canon

The question, "What is literature?" has exercised writers, critics and philosophers for a very long time. Places to start might be:

- Aristotle's *Poetics* is short and straightforward. Try it; you'll be surprised.
- Plato, *The Republic*, Books 2, 3 and 10. This is one of the earliest and most influential discussions of literature and, in these sections, is not too hard or too long.

More recent attempts to answer the question include René Wellek and Austin Warren's *Theory of Literature* (3rd ed.) (Harmondsworth: Peregrine, 1963), where they outline their understanding of the issues. Derek Attridge's excellent and clear *The Singularity of Literature* (London: Routledge 2004) offers a very interesting and innovative account of the literary. Jonathan Culler outlines a very different answer in his *Structuralist Poetics: Structuralism, Linguistics and the Study of Literature* (London: Routledge, 1975). There is a challenging but fairly accessible discussion of literature by Jacques Derrida, one of the most influential contemporary thinkers, in an interview, "This Strange Institution Called Literature," in Jacques Derrida, *Acts of Literature*, edited by Derek Attridge (London: Routledge, 1992).

Again, the canon is a subject that has generated a great deal of controversy. In addition to the Eliot material already mentioned, this is a small selection of accessible books on the subject:

- Harold Bloom, *The Western Canon* (London: Macmillan, 1995).
- Alastair Fowler, *Kinds of Literature* (Oxford: Clarendon Press, 1982).

- John Guillory, *Cultural Capital* (London: University of Chicago Press, 1993).
- Robert von Hallberg (ed.), *Canons* (London: University of Chicago Press, 1984).
- Barbara Herrstein Smith, *Contingencies of Virtue* (London: Harvard University Press, 1988).
- Frank Kermode, *The Classic* (London: Harvard University Press, 1983).
- Ankhi Mukherjee, *What Is a Classic? Postcolonial Rewriting and Invention of the Canon* (Stanford, CA: Stanford University Press, 2013).
- Robert Scholes, *Textual Power* (London: Yale University Press, 1985).

A good anthology of writers on these issues is Lee Morrissey, *Debating the Canon: A Reader from Addison to Nafisi* (London: Palgrave Macmillan, 2005).

The citation from Chinweizu, Onwuchekwa Jemie and Ihechukwu Madubuike comes from *The Decolonization of African Literature* (Washington, DC: Howard University Press, 1983). Toni Morrison is quoted from "Unspeakable Things Unspoken: The Afro-American Presence in American Literature," *Michigan Quarterly Review* 27 (1) (1989): 1–34.

Henry Louis Gates, Jr.'s *Loose Canons: Notes from the Culture Wars* (Oxford: Oxford University Press, 1992) is a witty discussion of the canon and the so-called culture wars, and his comments and those of the editors are cited from *Contemporary African American Literature: The Living Canon*, edited by Lovalerie King and Shirley Moody-Turner (Bloomington: Indiana University Press, 2013). The 1978 anthology I mention is *America in Literature*, edited by David Levin (New York: Wiley, 1978).

7: Castle Shakespeare

Here is just a very small selection of books relevant to the debates outlined in this chapter:

- Jonathan Bate, *The Genius of Shakespeare* (London: Picador, 1997).
- Jonathan Dollimore and Alan Sinfield (eds.), *Political Shakespeare: New Essays in Cultural Materialism* (Manchester: Manchester University Press, 1985).
- Malcolm Evans, *Signifying Nothing: Truth's True Contents in Shakespeare's Texts* (Harvester: Brighton, 1986).
- Terence Hawkes, *That Shakespeherian Rag: Essays on a Critical Process* (London: Methuen, 1986).

FURTHER READING

- Terence Hawkes, *Meaning by Shakespeare* (London: Routledge, 1992).
- Graham Holderness (ed.), *The Shakespeare Myth* (Manchester: Manchester University Press, 1988). This includes David Hornbrook's article, "'Go Play, Boy, Play': Shakespeare and Educational Drama."
- Sean McEvoy, *Shakespeare: The Basics* (London: Routledge, 2000).
- Kiernan Ryan, *Shakespeare* (3rd ed.) (Basingstoke: Palgrave, 2002).
- Gary Taylor, *Reinventing Shakespeare: A Cultural History from the Restoration to the Present* (London: Hogarth Press, 1989).

The quotations from Fay Weldon are in *Letters to Alice, on First Reading Jane Austen* (New York: Basic Books, 1999), pp. 11–20). Ludwig Wittgenstein discusses Shakespeare in *Culture and Value* (Oxford: Blackwell, 1998 edition). The article by John Yandell is "Reading Shakespeare, or Ways with Will," in *Changing English* 4 (2) (1997): 277–294.

On the subjects of new historicists and presentists, James Shapiro is the author of the widely celebrated *1599, A Year in the Life of William Shakespeare* (London: Faber & Faber, 2005). Shapiro is also the editor of a really superb anthology of writing called *Shakespeare in America: An Anthology from the Revolution to Now* (New York: Library of America, 2014). Among Stephen Greenblatt's many stylish and readable books are *Hamlet in Purgatory* (Princeton, NJ: Princeton University Press, 2001) and *Will in the World: How Shakespeare Became Shakespeare* (London: Jonathan Cape, 2004). An admirable introduction is *Stephen Greenblatt*, by Mark Robson (London: Routledge, 2008). Terence Hawkes edited a collection called *Shakespeare in the Present* (London: Routledge, 2002), as well as a series called "Accents on Shakespeare," from Routledge, that explores all sorts of new angles in this field. Ewan Fernie writes powerfully on "Shakespeare and the Prospect of Presentism," too, in *Writing about Shakespeare (Shakespeare Survey 58)*, edited by Peter Holland (Cambridge: Cambridge University Press, 2005). I've quoted from both of these. Fernie, too, with Simon Palfrey, edits a series of short, accessible and polemical books called *Shakespeare Now!* (New York: Bloomsbury).

Ayanna Thompson's book, *Passing Strange: Shakespeare, Race, and Contemporary America* (New York: Oxford University Press, 2011), is powerful and accessible. Emma Smith is the author of *The Cambridge Shakespeare Guide* (Cambridge: Cambridge University Press, 2012) and, with Laurie Maguire, of the highly entertaining and informative *30 Great Myths about Shakespeare* (Oxford: Wiley-Blackwell, 2013). Good recent discussions of adaptations are in Alexa Huang and Elizabeth Rivlin (eds.), *Shakespeare and the Ethics of Appropriation* (New York: Palgrave Macmillan, 2014) and Lukas Erne's *Shakespeare as Literary Dramatist* (Cambridge: Cambridge University Press, 2013). The now infamous business book is by Norman Augustine and Kenneth Adelman, *Shakespeare in Charge: The Bard's Guide to Leading and Succeeding on the Business Stage* (New York: Hyperion-Talk-Miramax, 1999). Jerry Martin's *The Shakespeare File: What English Majors Are Really Studying: A Report by the American Council of Trustees and Alumni* (Washington, DC: National Alumni Forum, 1996) is a polemical account of some of the issues I have discussed.

It would be fair to say that writing on Shakespeare has led to some of the most beautiful and interesting critical prose. A really good example is the very moving and perceptive introduction by Jonathan Bate to an edition of Shakespeare's *Complete Works: The RSC Shakespeare*, edited by Jonathan Bate and Eric Rasmussen (Basingstoke: Macmillan, 2007).

8: The author is dead?

The essays to which I refer are Roland Barthes's "The Death of the Author" (equally interesting and perhaps more useful is his essay "From Work to Text"), Michel Foucault's "What Is an Author?" and W. K. Wimsatt and M. C. Beardsley's, "The Intentional Fallacy." All these and a great deal more relevant material are included in an excellent reader edited by Seán Burke, *Authorship from Plato to the Postmodern* (Edinburgh: Edinburgh University Press, 1995). Burke has also written two very good studies of this issue, critical of the death of the author idea: *The Death and Return of the Author* (Edinburgh: Edinburgh University Press, 1992) and *The Ethics of Writing* (Edinburgh: Edinburgh University Press, 2008). An outstanding, if demanding account of these issues of authorial intention is Kaye Mitchell, *Intention and Text* (London: Continuum, 2008). Andrew Bennett discusses the author interestingly and clearly in *The Author* (London: Routledge, 2005).

9: Metaphors and figures of speech

Perhaps the most significant and accessible recent work on metaphor has been carried out by George Lakoff and Mark Turner, sometimes in collaboration. Many of the ideas and examples in this chapter are drawn from their work. Especially good and highly recommended is *More Than Cool Reason: A Field Guide to Poetic Metaphor* (London: University of Chicago Press, 1989). Other works include George Lakoff and Mark Johnson, *Metaphors We Live By* (London: University of Chicago Press, 1980) and George Lakoff's

Women, Fire and Dangerous Things: What Categories Reveal about the Mind (London: University of Chicago Press, 1987). Also interesting on this is a reader edited by Deborah Cameron, *The Feminist Critique of Language* (London: Routledge, 1998 edition). More complex but very rewarding is Jacques Derrida's essay, "White Mythology: Metaphor in the Text of Philosophy," in *Margins of Philosophy* (London: Harvester, 1982). Slightly at a tangent to all this but still about the power of language is J. L. Austin, *How to Do Things with Words* (Oxford: Clarendon Press, 1962). This is a very famous and influential account of language use. At the beginning, he writes, "What I shall have to say here is neither difficult nor contentious," and – amazingly – that's (mostly) true.

10: Narrative and closure

There is a great deal of material on narrative and narratology but less on closure. A very good starting point is H. Porter Abbot's *Narrative* in the Cambridge Introduction series (Cambridge: Cambridge University Press, 2002). Shlomith Rimmon-Kenan's *Narrative Fiction: Contemporary Poetics* (London: Routledge, 1990) is exhaustive. An excellent selection of key texts, which covers previous developments like Propp, as well as new ideas on this topic, is *The Narrative Reader*, edited by Martin McQuillan (London: Routledge, 2000). The best example of so-called narratology in full flow is Gérard Genette, *Narrative Discourse*, translated by Jane E. Leavis (Oxford: Blackwell, 1980). A more demanding but worthwhile book on narrative is Peter Brooks, *Reading for the Plot: Design and Intention in Narrative* (London: Harvard University Press, 1992). Frank Kermode's oblique, beautiful, and significant book on closure, *The Sense of an Ending*, has been reprinted recently (Oxford: Oxford University Press, 2000).

11: Creative and critical rewriting

There is a growing list of books that explore or argue over creative writing and its role.

D. G. Myers' *The Elephants Teach: Creative Writing Since 1880* (Chicago: University of Chicago Press, 2006) is a rewarding history of the history of creative writing, mainly in the United States. The title comes from a remark made when the novelist Vladimir Nabakov (1899–1977) was offered a chair (a professorship) in literature at Harvard: the famous linguist Roman Jacobson (1896–1982) asked, "What next? Shall we appoint elephants to teach zoology?" Mark McGurl's *The Program Era: Postwar Fiction and the Rise of Creative Writing* (Cambridge, MA, and London: Harvard University Press, 2009) is another history.

Ben Knights and Chris Thurgar-Dawson's *Active Reading: Transformative Writing in Literary Studies* (London: Continuum, 2006) is an insightful book, from which I have drawn much. It offers a clear argument about the importance of innovative practice in teaching English, along with an array of examples and case studies. Its aim is to fill the "woeful gap between the sublime theoretical possibilities and the actual teaching practice" in English that was analyzed by Rob Pope in his 1995 *Textual Intervention: Critical and Creative Strategies for Literary Studies* (London: Routledge, 1995).

Paul Dawson's *Creative Writing and the New Humanities* (London: Routledge, 2005) is also a history and an argument about the subject, concluding that the creative writer is in a position to be a radical public intellectual.

Angela Carter's comments come from *Shaking a Leg*, edited by Jenny Uglow (London: Chatto and Windus, 1997), one of the best and most interesting collections of nonfiction by a twentieth-century British novelist; it is a joy simply to browse through. The fantastic Facebook *Pride and Prejudice* is at http://www.much-ado.net/austenbook/. Douglas Cowie is the author of *Owen Noone and the Marauder* (New York: Bloomsbury, 2005), *Tin Pan Alley* (Orange, CA: Black Hill Press, 2013) and *Away, You Rolling River* (Orange, CA: Black Hill Press, 2014). Geoffrey Hartman's words come from his *Criticism in the Wilderness* (New Haven, CT: Yale University Press, 1980): his own creativity is described in his autobiography, *A Scholar's Tale* (New York: Fordham University Press, 2007).

For a different, perhaps more amusingly cynical view of English and creative writing, you might want to look at Julie Schumacher's *Dear Committee Members* (New York: Doubleday, 2014).

12: English, identity and politics

This is a huge and growing area. One of the most important books here is Benedict Anderson, *Imagined Communities: Reflections on the Origin and Spread of Nationalism*, revised and extended edition (London: Verso, 1991). The work of Homi Bhabha is often hard but very rewarding; see his edited collection *Nation and Narration* (London: Routledge, 1990) or his more challenging essays, *The Location of Culture* (London: Routledge, 1994). I quote from an essay called "DissemiNation: Time, Narrative and the Margins of the Modern Nation," which appears in both these. Salman Rushdie's reflections in *Imaginary Homelands: Essays and Criticism 1981–1991* (London:

Granta Books in association with Penguin, 1991) are also very illuminating (the citation from Rushdie is from an essay called "In Good Faith"). For linked accounts of similar issues, see the groundbreaking work by the critic Edward Said, including his justly celebrated *Orientalism* (London: Pantheon Books, 1978) and its follow-up, *Culture and Imperialism* (London: Vintage, 1994). Edward Said also writes very interestingly on politics, especially in his collection, *The World, the Text and the Critic* (London: Faber & Faber, 1984). Also very illuminating in this area is the work of Paul Gilroy, especially his *Against Race: Imagining Political Culture beyond the Color Line* (Cambridge, MA: Harvard University Press, 2000), which refuses to duck any of the hard issues that these debates generate, and *After Empire* (London: Routledge, 2004). The citations from Robert Young come from his exhaustive *Postcolonialism: An Historical Introduction* (Oxford: Blackwell, 2001). Henry Louis Gates develops his ideas in his edited volume *Black Literature and Literary Theory* (London: Methuen, 1984) and in detail in *The Signifying Monkey: A Theory of Afro-American Literary Criticism* (Oxford: Oxford University Press, 1988). Other writers discussing the same issues are Cornel West [there is a good *Cornel West Reader* (New York: Basic Civitas Books, 1999)] and bell hooks, whose work is very accessible. Her groundbreaking volume is *Ain't I a Woman: Black Women and Feminism* (London: Pluto, 1982), and, of her later collections of essays, *Teaching to Transgress: Education as the Practice of Freedom* (London: Routledge, 1994) is a rewarding read in this context. Raymond Williams is cited from *Keywords: A Vocabulary of Culture and Society* (London: Croom Helm, 1976).

Royall Tyler's *The Contrast* is available online.

13: Why study English?

There is a really superb online collection of articles on the state of the humanities at andreakastontange.wordpress.com/soh-articles/. You can find both Paul Jay and Gerald Graff's "Fear of Being Useful" (2012) and Michael Bérubé's "The Humanities, Declining?" (2013). I've already mentioned Geoffrey Galt Harpham's *The Humanities and the Dream of America* (Chicago: University of Chicago Press, 2011) and Louis Menard's *The Marketplace of Ideas* (New York: W. W. Norton, 2010), both of which are very constructive and interesting.

I've also drawn on Helen Small's *The Value of the Humanities* (Oxford: Oxford University Press, 2013) and on Martha Nussbaum's *Not for Profit: Why Democracy Needs the Humanities* (Princeton, NJ: Princeton University Press, 2012).

The UK government's Quality Assurance Agency document, "Subject Benchmark Statement for English," is at http://www.qaa.ac.uk/en/Publications/Documents/SBS-English-15.pdf.

Charles William Eliot also wrote, "To have been a schoolmaster or college professor thirty years only too often makes a man an unsafe witness in matters of education: there are flanges on his mental wheels which will only fit one gauge."

Conclusion

Harold Rosen's lecture, in *Neither Bleak House nor Liberty Hall: English in the Curriculum* (London: Institute of Education, 1981), is a bit dated now but a sterling statement of principles. Hannah Arendt is cited from *The Human Condition* (London: University of Chicago Press, 1998). Simon Swift's *Hannah Arendt* (London: Routledge, 2008) is a great introduction to her work for literature students.

Index

Abbot, H. Porter 157
Abrams, M. H. 151
accountancy 137
Achilles 97
Adams, John Quincy 71
Addams, Jane 24
Adelman, Ken 69, 156
Alcott, Louisa May 92
American Council of Trustees and Alumni 69
analepsis 105
Anderson, Benedict 122, 158
Angelou, Maya 70
Angry Birds 134
Animal Farm (Orwell) 96
animism 96
Anthology 57
anthropomorphism 96
Arendt, Hannah 147, 160
Aristotle 39, 153
Arnold, Matthew 16
Atherton, Carol 150
Attridge, Derek 153
Atwood, Margaret 92
Augustine, Norman 69, 156
Austen, Jane 20, 86, 89, 106, 114
Austin, J. L. 157

authorial intention 83–94
Ayres, David 151

Bacon, Francis 12
Baldick, Chris 150
Baldwin, James 127
Barry, Peter 151
Barthes, Roland 84, 89, 93, 156
Bate Jonathan 67, 72, 75, 154, 156
Batman 105
Beardsley, Monroe 84, 87, 88, 156
Beaumont, Francis 71
Belles Lettres 15
Bennett, Andrew 151
Bentham, Jeremy 134
Bertens, Hans 151
Bérubé, Michael 141, 159
Bhabha, Homi 123, 124, 158
Bikini Bottom 105
biography 13, 47, 48,
biology 7
Blake, William 57
Bleiman, Barbara 38
Bloom, Harold 60, 153
Breaking Bad 72
Briar Rose 104
Bristol, Michael 68

INDEX

Brooks, Cleanth 19
Brooks, Peter 157
Brown, James 98
Burke, Seán 156
Burney, Frances 89
Burns, Robert 98

Cain, William 151
Caliban 75
Cameron, Deborah 157
canon 20, 25, 53–63
Capitol Hill 66, 126
Cars 96
Carter, Angela 113, 158
Césaire, Aimé 75
Chaucer 20, 88, 90
chemistry 6, 7, 40
Child, Lydia Maria 60
Chinweizu 59, 153
Cinderella 104
civilizing mission 16, 20, 24, 25
Civil Rights Movement 5
Civil War 16, 127
Clancy, Tom 91
Clinton, Bill 67
close reading 45, 48
closure 103–10
Conrad, Joseph 105, 106
The Contrast (Tyler) 122
Council of Trent 55
Cowie, Doug 116, 158
Creative Writing Workshop 116
crisis of the humanities 133
critical rewriting 114, 117
"Criticism, Inc" (Ransom) 19
Cromwell, Oliver 71
Culler, Jonathan 151, 153
cultural heritage 131, 124
cultural materialist/materialism 70, 73, 75, 76, 79
Culture and Anarchy (Matthew Arnold) 16
culture wars 4, 30

Dante 99
Darwinism 92
Dawson, Paul 117, 158

dead metaphor 98
De Cervantes, Miguel 79
De Tocqueville, Alexis 71
De Vega, Lope 72
defamiliarization 98
Deleuze, Gilles 101
Democrats 126
Derrida, Jacques 102, 153, 157
Dickinson, Emily 124
diegetic 105
digital criticism 48
digital technology 8
digitization 47
disciplinary consciousness 33–42, 145, 153; as script 25
Dodson, Polly 116
Dollimore, Jonathan 154
Doyle, Bryan 150
DuBois, W.E.B. 127

Eagleton, Terry 151
East India Company 15
Egypt 12
Eliot T. S. 18, 58, 87, 152, 153
Eliot, Charles William 135, 160
Elsa 96
Emerson, Ralph Waldo 16, 56, 68
environment 7, 144
Erne, Lucas 78, 156
Evans, Malcolm 154
Evans, Maurice 73
extrinsic attitude 43, 46, 47, 48, 49, 128, 129, 132

Fables (Willingham) 104
Facebook 114
fairy tales 104, 105
Fault in Our Stars, The 66
feminism 7, 28, 39, 48, 144
Fernie, Ewan 78, 155
fictocriticism 117
Finding Dory 96
Finding Nemo 96
Finke, Laurie 151
first-person narrator 105, 106
Fitzgerald, F. Scott 41, 105

INDEX

Fletcher, John 71, 79
focalization 106
Folger Shakespeare Library 66
Ford 39
formalism 48
Foucault, Michael 92, 156
Fowler, Alastair 153
Frankenstein (Shelley) 39
Freud, Sigmund 47
Frost, Robert 20, 99
Frozen 96

G. I. Hamlet 74
Gardiner, Michael 150
Gates, Henry Louis 62, 128, 154, 159
Geiger, Roger 150
Genette, Gérard 105, 157
genre 55; conventions 56
geography 34, 36
geology 12
geometry 6
Gilroy, Paul 159
globalization 7, 144
Golden Treasury of English Verse (Palgrave) 56–7
Graff, Gerald 111, 138, 150, 159
Greek 13, 16, 16, 55, 58, 97
Greenblatt, Stephen 78, 117
Guillory, John 154

Habib, M.A.R. 151
Haley, Alex 127
Harpham, Geoffrey Galt 33, 150, 151, 159
Harry Potter 14
Hartley, L. P. 98
Hartman, Geoffrey 117, 158
Harvard University 135
Hattie, John 4, 42
Hawkes, Terence 74, 154, 155
Hawthorne, Nathaniel 60
Heart of Darkness (Conrad) 105, 106, 107
hermeneutics 28, 31
heuristic 111, 114
historicism 46, 48
history 7, 13, 26, 28, 35

Hobbit, The 136
Holderness, Graham 155
Holland, Peter 155
Holmes, Sherlock 12, 150
Homer 58
hooks, bell 159
Hornbrook, David 75, 155
Huang, Alexa 155
Hughes, Langston 127
hyperbole 96

identity politics 130
India 15
Industrial Revolution 90
intrinsic artistic worth 25, 26
Intrinsic attitude 43–8, 49, 128, 129, 132

Jacobson, Roman 157
James, Henry 20, 108, 123
Jargon 40
Jaschik, Scott 152
Jay, Paul 138, 159
Jemie, Onwuchekwa 59, 153
Johnson, Barbara 151
Johnson, Robert 99
Jonson, Ben 68, 71
Joyce, James 104

Kermode, Frank 108, 109, 154, 157
Kettell, Samuel 56
King, Lovalerie 62, 154
King, Stephen 91, 92
Knights, Ben 35, 114, 143, 152, 158

Lacan, Jacques 47
Lakoff, George 97, 99, 100, 156
Latin 14, 16
Leach, Susan 76
Leitch, Vincent 30, 151
Levin, Bernard 66
Library of Congress 66
linguistics 26
literary critics 17, 26, 34
literary theory 26, 28, 29, 30, 31, 34, 144; absence of a Grand Unifying Theory of Everything 28

163

London 45
Longfellow, Henry 16, 56
Lord of the Rings, The 136
Los Angeles 77
Lounsbury, Thomas 57

Macaulay, Thomas Babington 15
Macbeth 56, 135
MacIntyre, Alastair 152
Madubuike, Ihechukwu 5, 153
magic 13
Maguire, Laurie 155
Martin, Jerry 156
Marxism 92
mathematics 34, 36
May, Steve 112
McDonald, Ronan 152
McEvoy, Sean 155
McGowan, John 151
McGuffey, William Holmes 72
McGurl Mark 157
McQuillan, Martin 157
Melville, Herman 20, 92
Menand, Louis 18, 19, 134, 137, 150
metacognition 4, 41
metaphor 97–101
metonymy 96
Milton, John 56, 85, 88
mimetic 53
Minute on Indian Education 15
Mitchell, Kaye 156
Moby Dick 107
Moody-Turner, Shirley 62, 154
Moore, George 122
Moretti, Franco 39
Morrison, Toni 61, 154
Morrissey, Lee 154
Mosley, Walter 91
Mukherjee, Ankhi 154
Myers, D. G. 157

Nabakov, Vladimir 157
Napoleonic Wars 16
narrative 103–10
National Education Association 75
National Endowment for the Arts 68

national identity 121–3
New Criticism 18–24, 45, 41
New York 122
Nietzsche, Friedrich 101, 102
Norse 14
Nussbaum, Martha 136, 146, 159

Obama, Barak 127
objectification 40
Onwuchekwa, Jemie 59, 153
Ortberg, Mallory 114
Orwell, George 96
Othello 38

Palfrey, Simon 155
Palgrave, Francis Turner 56–7
palimpsest 12–13
Paradise Lost (Milton) 85, 88
pathetic fallacy 96
Pericles 97
philology, philologists 13, 14, 17, 19, 21, 30, 34, 55
philosophy 12, 26
Pirates of the Caribbean 15
Plato 153
Pope, Rob 158
practical criticism 45, 48
Pratchett, Terry 126
Prince Charming 104
prolepsis 105
Propp, Vladimir 104, 105
Prospero 75
psychoanalysis 47, 48
psychology 35
push-pin 134
Pynchon, Thomas 108

Quality Assurance Agency 138, 160

Ransom, John Crowe 18, 19
Rasmussen, Eric 156
reification 40
Renaissance 55–6
Representative Men (Emerson) 68
Republicans 126
rhetoric 14, 30

INDEX

Richards, I. A. 18, 19
Rimmon-Kenan, Shlomith 157
Rivkin, Julie 151
Rivlin, Elizabeth 155
Robson, Mark 155
Roman 13, 18, 55, 58
Romantic Movement 91
Rosen, Harold 15, 160
Royle, Nicholas 151
Rushdie, Salman 125, 158
Ryan, Kiernan 69, 155
Ryan, Michael 151

Said, Edward 159
Saxon 13
Scholes, Robert 154
Schumacher, Julie 158
science fiction fandom 41
Scott, Patrick 152
sexuality 7
Shakespeare, William 13, 16, 18, 38, 41, 63–79, 85, 92, 105, 117, 124, 129; *As You Like It* 68; *Cardenio* 79; *Hamlet* 67, 73, 114; *Henry V* 72; *Julius Caesar* 66, 78; *King Lear* 46, 68; *Othello* 85, 20, 46; *The Merchant of Venice* 69; *Romeo and Juliet* 66, 74; *The Taming of the Shrew* 66, 114; *The Tempest* 75
Shapiro, James 78, 155
Sharpley-Whiting, T. Denean 150
Shelley, Mary 38; *Frankenstein* 39
Sidney, Sir Phillip 56, 141
simile 97
Sinfield, Alan 154
Sir Gawain and the Green Knight 90
Small, Helen 135, 159
Smith, Barbara Herrstein 154
Smith, Emma 79, 155
Snow White 104
sociology 13, 40, 26
Sophocles 72
SparkNotes 37
Spenser, Edmund 56
spoiler alert 107
SpongeBob 105

Stedman, Edmund Clarence 56
Steele, Danielle 91
Stowe, Harriet Beecher 127
Swift, Simon 160
synecdoche 96

Tate, Allen 18
Taylor, Gary 71, 74, 155
Teleology 107
10 Things I Hate About You 66
Tetris 134
Theory of Literature (Wellek and Warren) 44
third-person Narrator 106
Thompson, Ayanna 77, 155
Thugar-Dawson, Chris 114, 143, 158
Thurber, James 56
Tillyard, E.M.W. 58
Tolkien, J.R.R. 136
Toward the Decolonisation of African Literature (Chinweizu, Jemie, and Madubuike) 59, 153
Toy Story 96
tradition 35, 36
"Tradition and the Individual Talent" (Eliot) 18, 58
Turner, James 150
Turner, Mark 97, 99, 100, 156
Tyler, Royall 122, 59
Tyson, Lois 151

Underwood, Ted 14, 150
University of London 14
University of Missouri 17
utilitarianism 134

Viswanathan, Gauri 150
von Hallberg, Robert 154

Walden Pond 124
Wallace, David Foster 149
Wandor, Michelene 113,
Warner, Michael 150
Warren, Austin 44; 153
Warton, Joseph 56
Washington D.C. 66

165

Webster, Lucy 28
Weldon, Fay 65, 66, 155; *Letters to Alice, on First Reading Jane Austen* 65
Wellek, René 44, 153, 154
Wellmon, Chad 152
West, Cornel 159
White House 96
Wikipedia 37
Williams, Jeffrey 151
Williams, John 17, *Stoner* 17
Williams, Raymond 123, 150, 159
Willingham, Bill 104
Will Power to Youth 77
Wimsatt, W. K. 84, 87, 88, 152
Wire, The 72
Wittgenstein, Ludwig 30, 73, 143, 151, 155
Wollstoncraft, Mary 13
Women's Studies 26
Words on the page 5, 18, 23, 40, 48, 49
Wordsworth, William 45
World War I
World War II 72, 73
Wright, Richard 127

X, Malcolm 127

Yandell, John 67, 155
Yeats, William Butler 134
Young, Robert 127, 159
YouTube 38, 61

Made in the USA
San Bernardino, CA
28 August 2018